CLIMATIC CHARACTERISTICS AND WATER BALANCE

CLIMATIC CHARACTERISTICS AND WATER BALANCE

(A Study of Uttar Pradesh)

LALITA DEVI

CONCEPT PUBLISHING COMPANY, NEW DELHI 110059

ISBN 81-7022-389-X

First Published 1992

Published and Printed by
Ashok Kumar Mittal
Concept Publishing Company
A/15-16, Commercial Block, Mohan Garden
NEW DELHI 110059 (India)

Lasertypeset by
Microtech Advance Printing Systems (Pvt.) Ltd.
H-13, Bali Nagar
NEW DELHI 110015

FOREWORD

THIS BOOK, entitled *Climatic Characteristics and Water Balance: A Study of Uttar Pradesh* is a new venture in geographical studies attempted by Dr. (Mrs.) Lalita Devi. The exercises necessary for the preparation of materials needed for this book were conducted by her during her Ph.D. research.

Uttar Pradesh, a prominent state of the Republic of India is well known for its vastness in size, density of population and varied nature of terrain. It is far away from the Bay of Bengal and enjoys high degree of continentality and consequently decreasing amount of precipitation during the southwest monsoon from east to west. It was thus quite a befitting area selected by the researcher to carry out her exercises with respect to determination of climatic water balance. For an agricultural state like Uttar Pradesh, where the mainstay of the people in general is farming, assured supply of moisture needed by the crops in different season becomes quite important. We are all well acquainted with the nature of the southwest monsoon. It cannot be relied upon in regard to the total amount of precipitation, its regional distribution, its effectiveness on the natural reserves of the ground-water etc. Therefore, for successful farming, adequate and dependable means of irrigation must be developed. For any sound economic and social and particularly agricultural planning, the seasonal deficiency and surplus of water in the soil must be known to the planners and policy-makers. As such there is a very sound case for a scientific diagnosis of the rainfall regime and water balance in the state. The exercises carried out by Dr. Lalita Devi in this book make an attempt in this direction. She has worked out the monthly climatic water balance in different parts of the state on the basis of the necessary climatic data available for 39 meteorological stations. On the basis of the soil and vegetation types she has also estimated the moisture holding capacity of the soil in different regions. For the calculation of water balance at different representative stations, she has used the

Thornthwaite's method and the relevant tables prepared by Thornthwaite and Mather. The calculations are based on the average climatic data spread over a number of years (10 years for some and 30 years for the rest). She has also worked out the water balance of typical rainy and drought years. In addition she has worked out the water balance of 14 selected stations for 16 consecutive years to observe the year to year changes. She has also done useful exercises in calculating climatic water balance of 22 stations using the value of Potential Evapotranspiration based on Penman's empirical law, just for a comparative assessment of the merits of Thornthwaite's and Penman's empirical laws.

I am sure the regional planning bodies will get lot of assistance from this study in their attempt to improve the irrigational facilities in different parts of the state. A proper utilization of the materials available in this book will also help the water management authorities in various parts of the state, especially in the drought prone areas. For future researchers in this field, this book will provide useful and scientific guidelines. The book will be one of prize possessions on the shelves of any serious researcher and planner on the subject.

Professor of Geography — DR. K.N. SINGH
B.H.U., Varanasi

PREFACE

UTTAR PRADESH (U.P.) is the new post-Independence name of the province, previously known as the United Provinces during the British regime, comprising the two old Subas of Agra and Oudh of the Mughal period. It touches the northern boundary of the Indian Republic and is the largest and most populous state of northern India. It covers a major portion of the fertile Ganga plain and spreads its arms northward over the Himalayan region and southward over the plateau region. It is the diversity of relief, vegetation, soil, climate, agricultural economy and society, which prompted the present investigator to choose this state for her research. A thorough assessment and analysis of climatic elements and climatic water balance will help in utilizing its hidden potentiality for the benefit of its large rural population.

Chapter 1 deals with the physical properties of the land, i.e., geology, physiography, drainage, soil and vegetation. Chapter 2 deals with the spatio-temporal variations in distribution of climatic elements, i.e., temperature, pressure and wind, relative humidity, rainfall and frequency of atmospheric disturbances (hail, thunder, fog, dust storm and squall). Climographs and hythergraphs have been drawn to assess their importance with respect to human comfort and suitability of climatic conditions for different crops. Seasonal distribution of these climatic elements have been assessed in Chapter 3. During the southwest monsoon the state receives 70 to 90 per cent of its annual rainfall. The position and intensity of the seasonal trough plays a major role in controlling the distribution of rainfall during this season. On the other hand, less than 10 per cent rainfall is received during the winter season by the western disturbances showing a decreasing trend from west to east. Cold waves are generally experienced from November to March and heat waves from April to July. In Chapter 4 climatic types have been recognised on the basis of Lang's Rain factor (1920), De Mortonne's Index of Aridity (1926), Koppen's Scheme (1936), Kira's Index of Moist and Dry Climate (1945) and Thornthwaite's Scheme

(1955). Among these, Thornthwaite's Scheme has gained popularity mainly because of two important factors, i.e., potential Evapotranspiration and Moisture Index which are considered in determining the types of climate. In Chapter 5 an attempt has been made to the microclimatological variations in a hypothetical urban area (a case study of Varanasi) as well as in mountainous areas (a case study of the U.P. Himalaya) with respect to temperature, relative humidity and rainfall.

Chapter 6 deals with the concept of Potential Evapotranspiration as well as the *modus operandi* of its estimation and spatial distribution. Average climatic water balance and its components, i.e., actual evapotranspiration, water surplus, water deficit, soil moisture utilization and soil moisture recharge as well as Humidity index, Aridity index, Moisture index and Moisture adequacy index have been assessed in Chapter 7. Thornthwaite's water balance components have also been compared with those of Penman's results. Year-wise water balance for a 16-year period (1957-72) of 14 selected stations have been computed in order to determine the trend in the pattern of water surplus and deficit as well as to recognize the excessive Rainy and Drought years in the last chapter (Chapter 8). It may be mentioned here that some of the materials of this thesis were contributed by the present authoress and have already appeared in various publications.

The present study has exposed a number of problems that the state faces today. An attempt has been made to diagnose the main problems particularly related to the spatio-temporal variations of water surplus and deficit. Relevant remedial measures have been suggested for solving these burning problems as they directly affect the agricultural prosperity of the state. If the Government's attention is drawn towards these problems and the suggested remedial measures are implemented, the state, no doubt, possesses a bright future prospects.

I offer my most sincere gratitude to my Supervisor Shri N. Prasad, retired Reader in the Department of Geography, Banaras Hindu University for his inspiration and able guidance. I am very much indebted to my teachers Prof. N.K.P. Sinha, Prof. K.N. Singh, Prof. H.H. Singh, Prof. R.N. Mathur Dr. V.R. Singh and Shri N.K. Singh for their valuable suggestions. I acknowledge my indebtedness to Prof. V.P. Subrahmanyam, Andhra University, Waltair, who helped me in solving the problems which cropped in during the course of the present work.

I owe a lot to the officials of the Regional Meteorological Centre,

New Delhi, India Meteorological Department, Pune and Lucknow Meteorological Centre for their co-operation and help.

I am highly grateful to my elder brother Shri Sharda Prasad, Chief Geophysicist in O.N.G.C. for his valuable company in the course of my data collection trips and my husband Dr. R.P. Verma, Private Physician for his moral support. I shall be failing in my duty if I don't thank Shri Om Prakash Srivastava, who extended Cartographical help. Lastly, I acknowledge with gratitude the financial assistance extended to me by C.S.I.R. during research period.

Varanasi LALITA DEVI

CONTENTS

LIST OF FIGURES

LIST OF TABLES

INTRODUCTION

DURING THE past few decades considerable attention has been paid to the study of climatic basis for the distribution of plants and soils. Plant ecologists, soil scientists and geographers have recognised that the development of such studies on a truly scientific basis is dependent on the qualitative and quantitative consideration of climate. They generally take two main climatic elements into consideration, temperature and precipitation, which are most decisive in determining the vegetal cover of the landscape.

Koppen first identified the boundaries of climatic regions from the studies of distribution of vegetation and then selected numerical climatic values for such boundaries. Though, his classification was simple for application and was widely used by scientists of the world, its boundaries were arbitrary and unscientific. Index of Aridity was derived by De Mortonne (1926) which was a slight modification of Lang's Rain factor (1920)[1] and had been widely used by geographers. Later on, Thornthwaite's Rational classification (1948)[2] and subsequently modified in 1955[3] was considered far better than the Koppen's (1936)[4] and Thornthwaite's (1931)[5] schemes.

Thornthwaite's Rational classification of climate introduced a new concept of 'Potential evapotranspiration' and claimed that climatic boundaries should be based on 'Potential evapotranspiration' and 'Moisture index'. Water surplus and deficit were computed from 'Climatic water balance'. Seasonal and regional patterns of positive and negative water balance exhibit a clear picture of water surplus and water deficit patterns.

V.P. Subrahmanyam brought out a climatic divisions of India in 1956[6] based on Thornthwaite's scheme of 1948. He found four climatic divisions in U.P., viz., Semiarid, Subhumid, Humid and Perhumid. He again classified India into climatic divisions in 1965[7] on the basis of the final scheme of Koppen and that of Thornthwaite. This time he exposed the presence of a number of smaller climatic regions in the state.

For the purpose of the present study, average rainfall and number of rainy days for a 50-year period for 200 stations and average temperature rainfall and relative humidity data of 39 stations (the average data are not for the same period for all stations) have been utilised. The computed values of 'Potential evapotranspiration' for 22 stations by Penman's method have been incorporated.[8] The water balance of 17 stations on the basis of Penman's P.E. have been calculated. Year-wise water balance for a 16-year period (1957-72) of 14 selected stations have also been computed. But the computed tables and their related graphs are not being included in the present study because it will unnecessarily increase the bulk of the thesis. A critical analysis of the tables has been given at the proper place in the text. The water balance graphs of all these stations for excessive Rainy and Drought years have been included in the text and their corresponding tables have been appended in Appendices H and I.

The necessary data for the present study have been taken from 'Monthly and Annual Rainfall and Number of Rainy days'.[9] 'Climatological Tables of Observatories of India'[10] and 'Monthly Weather Report'.[11] Some necessary data were collected personally from the Meteorological Offices at New Delhi and Pune. The average rainfall data for the periods 1901-1950, 1931-1960 and 1957-72 slightly differ from each other which posed a problem in this study, but necessary data for computing average climatic water balances were to be obtained and such data are available only in the Climatological Tables.

The soil types map has been drawn based on the 'Agricultural Atlas of Uttar Pradesh'.[12] On the basis of these soil types and the existing vegetation, the 'Moisture holding capacity' in different parts of the state were assessed in consultation with V.P. Subrahmanyam and tabulated in the text (Table 7.1). The 'Moisture holding capacity' varies from 5.0 to 40.0 cm in the State. In the present study, water balance has been computed incorporating these 'Moisture holding capacities'. The *modus operandi* followed by the state is the same as suggested by Thornthwaite and Mather in 'Instructions and Tables for Computing Potential Evapotranspiration and Water Balance'.[13]

Uttar Pradesh is primarily an agricultural state and it often faces the problems of floods and droughts. In the Per-humid and Humid regions of the U.P. Himalaya, considerable amount of water surplus exists in some months. The sub-humid regions of the *Tarai*, the central and the eastern U.P., have much variations of water surplus and deficit from year to year and hence are more liable to floods and droughts. In the semi-arid region of western U.P. water surplus is small or negligible, hence the soil of this region plays very little role as a moisture storage. Agricultural yields are largely dependent on the availability of adequate amount of moisture at appropriate times of crop growth. A knowledge of distribution of water deficit in space and time helps the experts in proper crop planning and the farmer to adjust his agricultural operations suitably. Similarly, the knowledge of water surplus provides a rough idea of stream flow and ground water reserves. These information are useful to the hydrologists and water management authorities in the design and operation of reservoirs, river projects and utilisation of groundwater storage. Besides, its use in proper water management, water surplus information helps the farmer in locating areas and periods of dependable water supply for successful crop raising.

NOTES

1. Lang, R.: Verwitterung und Bodenbildung als Einfuhrung in die Bodenkunde Stuttgart (1920).
2. Thornthwaite, C.W. : An Approach Toward a Rational Classification of Climate, *Geographical Review*, Vol. 38, New York (1948), pp. 55-94.
3. Thornthwaite, C.W. and Mather, J.R.: The Water Balance, *Publication in Climatology*, Vol. 8, No.1, Centerton, New Jersey (1955).
4. Koppen, W. and Geiger, R.: *Das geographische system der Klimate*, Hundbuch der Klimatologie 1, C (1936).
5. Thornthwaite, C.W.: The Climates of North America According to a New Classification, *Geographical Review*, Vol. 21, New York (1931), pp. 633-655.
6. Subrahmanyam, V.P.: Climatic Types of India According to Rational Classification of Thornthwaite, *Indian Journal of Meteorology and Geophysics*, Vol. 10, No.1, New Delhi (1956), pp. 253-259.
7. Subrahmanyam et al.: Koppen and Thornthwaite System of Climatic Classification as Applied to India, *Annals of Arid Zone*, Vol. 4, No. 1, Jodhpur (1963), pp. 45-55.
8. Rao, K.N. et al. : Potential Evapotranspiration (PE) Over India, *Scientific Report* 136, I.M.D. (1971).
9. Koteswaram, P. (Ed.): Monthly and Annual Rainfall Number of Rainy Days, Period 1901-1950, Part III, A.I.M.D. (1969).
10. Government of India: *Climatological Tables of Observatories of India 1931-1960*, I.M.D. (1967), pp. 85-118; 349-364.
11. Government of India: *India Weather Review*, Monthly Weather Report, I.M.D., Manager of Publications, Delhi.
12. Tewari, S.B.: *Agricultural Atlas of Uttar Pradesh*, Pantnagar, India (1971).
13. Thornthwaite, C.W. and Mather, J.R.: Instructions and Tables for Computing Potential Evapotranspiration and Water Balance, *Publication in Climatology*, Vol. 10, No. 3, Centerton, New Jersey (1957).

1

PHYSICAL SETTING

LOCATION AND EXTENT

UTTAR PRADESH extends from 23° 52′ N 31° 28′ N and from 77° 04′ E to 84° 38′ E embracing a total area of 2,93,729 km², which amounts to about one-tenth (9.65 per cent) of the national total. It is bounded by Tibet and Nepal in the north, Himachal Pradesh in the northwest, Delhi and Haryana in the west, Rajasthan in the southwest, Madhya Pradesh in the south and Bihar in the east (Fig. 1.1). The Yamuna river forms a part of the western boundary and the Ganga, the Gandak and the Karmanasa, parts of the eastern boundary. Other boundaries are artificial. Its terrestrial extent can well be perceived by the fact, that even on equinoxes a vertical pole casts into midnoon shadow 38 per cent longer at its northern extreme than the southern and the local time at the eastern extreme is 38 minutes 16 seconds ahead of that at the western one. Its altitudinal variation too is quite immense varying from 46 m in the Ganga valley to 7,618 m in the U.P. Himalaya.

GEOLOGY

Uttar Pradesh covers parts of the three major structural divisions of India, viz., the folded Himalaya, the Ganga trough and the stable Deccan plateau.

The U.P. Himalaya

On the basis of various studies, the region falls under three broad stratigraphical zones: (a) the outer or sub-Himalayan zone, composed of sediments mostly of Tertiary age, (b) the Central or Lower Himalayan zone, composed of granite and other crystalline rocks of unfossiliferous sediments, and (c) the Higher Himalayan zone, composed of a series of highly fossiliferous sediments. The foothill

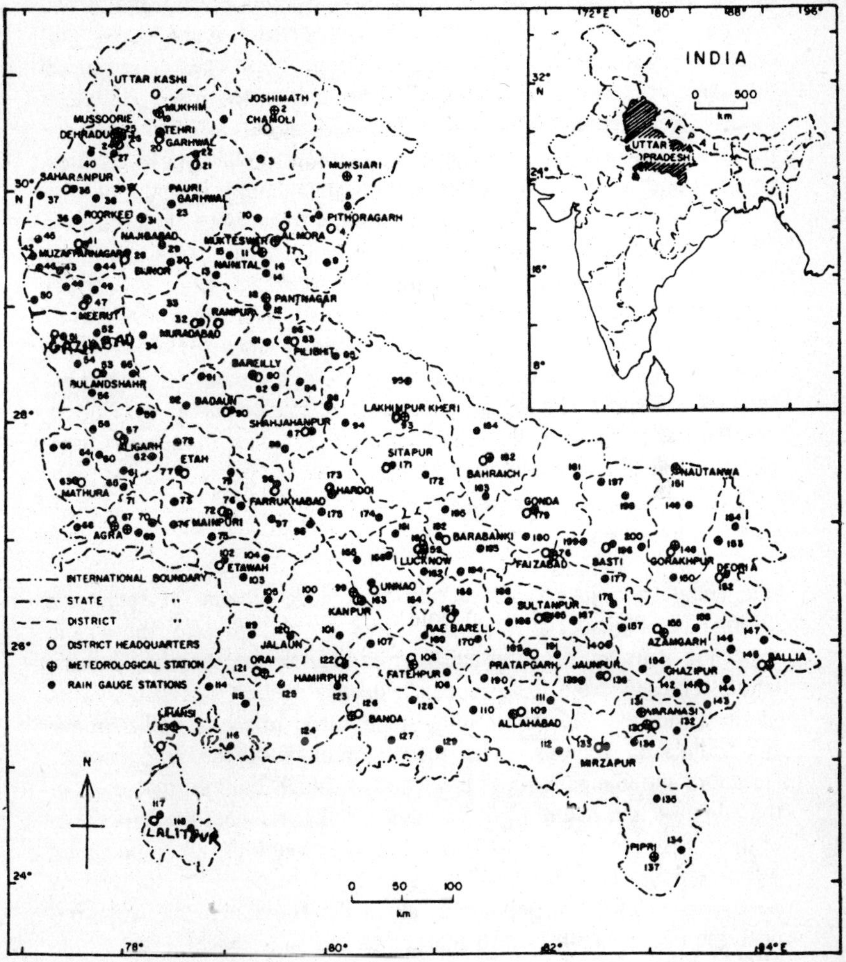

FIG. 1.1 Location Map: U.P. in India; Rain Gauge and Meterological Stations in U.P.

belt of the sub-Himalayan zone is built entirely of Siwalik sediments. Siwalik constitutes a great thickness of detrital rocks, clays and conglomerates measuring 5,000 to 5,500 m. These beds of sandstone and shale are separated from the Eocene beds of the lower Himalaya by the 'Great Boundary Falt'.[1] The Eocene beds of lower Himalaya are mostly composed of crystalline and metamorphic rocks-granites, gneisses and schists with unfossiliferous sedimentary deposits of very ancient *(Purana)* age. The main structural features of this zone are Krol belt, Deoban Tejam belt and Almora Dudatoli crystalline thrust sheet. The Higher Himalayan zone is separated from the lower Himalayan zone by the 'Main central thrust'. The innermost line of high ranges rises above the limit of perpetual snow (6,000 m). This zone is composed of a continuous series of highly fossiliferous marine sedimentary rocks ranging from the earliest Palaeozoic to the Eocene age. The main rocks of this zone are quartzites, magmatites, gneisses, garnet-schists, dioritic amphibolites etc.

The Ganga Trough

The Ganga and its tributaries carried a huge amount of sediments during the Pleistocene and sub-recent to the present times and deposited it to several thousand metres over the old basement rocks, which represent from the Archean *(Purana)* to the Gondwana system of the Peninsula and from Numulitic, Murrie to the Siwaliks of the sub-Himalayan regions.[2] The recent findings of the aeromagnetic survey of the Ganga Valley[3] have thrown much light on the depth of the alluvium and nature of the trough. It rather replaces the hypothesis of the frontal plain by that of downwarping, though the alluvium filling on an average is 1,300-1,400 m, decreasing gradually southward. A zone of over 8,000 m depth runs along the Himalaya. The westward decrease of the depth is well marked as the small basin, lying to the west of the Moradabad-Kathgodam alignment, has a depth of less than 6,000 m. The Doab, south of Bulandshahar in general, has less than 1,500 m alluvium depth. The alluvium depth of probably less than 3,000 m, occasionally found at places, covers the detached parts of the Peninsular block. The constituents of lethology can be classified under two subdivisions: (a) recent alluvium-*khadar*, the silt clay and sand particles renewed annually and occupying the riverine tracts (flood plains), and (b) older alluvium-*bhangar*, occupying the interfluvial zones above the general flood limits, the constituents experiencing slow and secular changes. The gravel of the Tarai and the Bhabar zones are the loosely set sediments ranging from fine silt and clay particles to coarse sands, pebbles and sometimes even boulders spread haphazardly in the zone of the break of slope at the foot of the Siwalik.

The Southern Stable Plateau

Parts of its northern fringe appear in Mirzapur, Banda and Jhansi districts. The Mirzapur upland can be further classified into three different geological formations: (a) the sedimentary rocks of the Vindhyan formation, north of the Son gorge, (b) the Sonpar country of metamorphics of different origin in the south, and (c) a narrow zone of the Son gorge separating the two zones. The Vindhyan formations are mainly Rohtas limestone, Bijaigarh shale and Kaimur sandstone. Quartzite cappings, however, are visible in isolated and precipitous crags overlooking the Son. Most of the Sonpar region is composed of metamorphic rocks like quartzite, gneiss, slate, phyllite micaschist and other complex rock formations almost all of them being crystalline in nature.

The three main geological systems represented in Bundelkhand are: (a) The Archean system, (b) The Vindhyan system, and (c) The recent deposits. The Archean system is represented by the Bundelkhand massif which is largely composed of crystalline igneous and metamorphic rocks. In geological literature, the typical rocks of this region is popularly termed as Bundelkhand gneiss. In fact, granite and gneiss are both conspicuous though the former predominates. The Vindhyan system forms a series of imposing escarpments of massive sandstone and limestone. In the northwest and northeast the system is covered by Ganga-Yamuna alluvium and in the southwest by the Deccan Trap. The recent deposits in Bundelkhand are represented by large scale alluvial deposits in the form of an 'embayment' into the granite country. The alluvial sediments are of fluviatile and subaerial formations of sand, silt and clay.

PHYSIOGRAPHY

Physiographically the state can be divided into four regions:

Mountainous Part of the Himalayan Ranges and Valleys

This region is popularly known as the U.P. Himalaya. It is bounded by the Tons in the west and the Kali in the east, which separate this area from the Himachal Pradesh and Nepal respectively. It covers about 38,000 km² and contains the three main Himalayan ranges: (a) Higher or Greater Himalaya, (b) Central or Lesser Himalayan, and (c) The Siwalik or Outer Himalaya.

(a) *The Greater Himalaya* in the state covers a part of the Zanskar range (altitude ranging between 4,800 and 6,000 m). About 6,600 km² of this zone remains perpetually snowclad. There are four series of glacier garlanded peaks: (1) Bunderpunch (6,315 m); (ii) Gangotri (6,614 m), Kedarnath (6,940 m), Badrinath (7,138 m); (iii) Kamet (7,756 m); and (iv) Nanda Devi (7,817 m), Dunagiri (7,066 m), Trisul (7,120 m) and Nanda Kot (6,816 m). These peaks are separated by transverse gorges of the Bhagirathi, the Vishnuganga, west Dhauliganga, east Dhauliganga and the Goriganga. Mala, Niti, Kungri-bingri and Lipu are the main passes between Tibet and U.P. The climatological significance of this mountain is that it acts as a barrier for the continental air mass from the north and the southwest monsoon from the south.

(b) *The Lesser Himalaya* covers a zone of 60 to 100 km wide in-between the Siwalik and the Greater Himalaya. On an average the elevation ranges from 1,500 m in the valley beds to 2,700 m along the crest of the ridges. There are two ranges - the Mussoorie and Nag Tibba, stretching almost parallel to each other. The whole zone is a tangled mass of ridges separated by deep valleys. The Nainital lake region, lying at its outer fringe and extending for 25 km in length and 4 km in width, has its own characteristic features. Other lakes are Bhim tal, Naukuchiya tal, Sal tal, Pama tal etc., which lie to the east of the Nainital lake and form a considerably low-lying open lake basin. Another important lake is Diuri tal lying 10 km to the northeast of Ukhimath.

(c) *The Siwalik* is a long and narrow chain of low hills trending northwest-southeast following the general trend of the Himalayas. The elevation ranges from 350 to 1,500 m. On the southern slopes, it has steep scarps while on the north it descends gently to the flat floored structural valleys called 'Duns'. Dehra Dun is the biggest, about 35 km long and 25 km wide. Other Duns are Kohtri, Chaukhamba, Patti and Kota.

The Narrow Tract Between the Ganga Plain and the Siwaliks

Immediately below the Siwalik range and separating it from the Ganga plain there lie two successive narrow elongated belts. Firstly the Bhabar a 'dry' boulder-strewn tract, usually covered with forests but remarkable for the virtual absence of surface water and secondly the Tarai a marshy stretch infested with reeds, tall grasses and scrubs. The origin of this region is like that of a piedmont alluvial plain. Topographically it is a low lying plain with slight undulations. On micro level, the Tarai reveals entirely a criss-crossed surface by small streams, nalas and swamps. The interfluves, river levees, and alluvial terraces provide suitable sites for human occupance.

The Ganga Plain

This tract occupies about three-fourths of the state. Its northern limit is marked by the Bhabar and Tarai tract and most of the southern limit by an irregular and broken margin of the rugged southern plateau. It represents an smooth alluvial tract. Drainage lines are the most conspicuous element of physiography (relief) in the plain, otherwise the entire plain is seemingly featureless, with imperceptively varying elevation of 60 m in the southeast to 275 m in the northwest. It is broken by river bluffs, ox-bow lakes, bad lands and ravines. A more perceptible relief is marked by the detached hillocks, cut off from the Vindhyan plateau in the southern part, where small flat topped and heavily eroded hillocks separated by broad embayment of alluvial flats are formed by aggradational action of the numerous rivers debauching with heavy loads and aggrading the surface. On the basis of micro physiographic units, this tract may be divided into the following subregions:

The Ganga-Yamuna Doab

It lies in the fork of the two rivers, the Ganga and the Yamuna, with its elevation varying between 90 and 275 m. It has three distinct features: (i) ravine land in the lower Yamuna plain, (ii) *bhangar* land stretching over the central part of the Doab, and (iii) low sandy ridges (*dhus*) mostly in the northern part. Patches of *Usar* (alkaline soils) are common in the lower and middle Doab.

The Ganga-Ghaghra Doab

The Ganga-Ghaghra Doab, lying north and east of the Yamuna-Ganga Doab, ranges in elevation from 274 m in the west to 48 m in the east. The regional slope is towards the southeast and the Ramganga, *Gomati* and Sarda flow in that direction. There are a number of *tals*, ox-bow lakes, deserted channels of rivers particularly along the Ghaghra, and local eminences such as river levees and bluffs.

The Saryu-Par Plain

It lies north of the Ghaghra river. The Rapti, the Burhi Rapti, the Chhoti Gandak etc. drain this plain. The *bhangar* lands have a series of sandy ridges known as *dhus* in this part. The *khadar* lands are occupied by numerous bluffs and pools and mark somewhat distinguishing features through numerous deserted channels and lakes in this broad *khadar* plain.

The Ganga-Par Plain

It is popularly known as the Trans-Ganga plain and is bounded by the Vindhyan upland in the south and the Ganga in the north. This region is confined between the serpentine channel of the Ganga and the rugged irregular edge of the Deccan foreland. It is characterised with low and small knolls, irregularly plunged notches, broken spurs, gullies and detached hillocks.

The Trans-Yamuna Plain

It lies between the Yamuna and the southern upland from Mathura and Allahabad. A considerable proportion of this region is broken by ravines of the Yamuna, Chambal, Betwa and Ken rivers. Tiny knolls and detached hillocks appear in the Mathura district. Eastwards, there appears spurs and knolls of Vindhyan sandstone as well as irregular notches in its edge.

The Southern Uplands

Two small tracts, which differ considerably from the plain, lie in the southwest and southeast of the state. The western one is a part of the

Bundelkhand upland whereas the eastern one forms a part of the East Vindhyas. The general elevation ranges from 150 to 350 m.

The Bundelkhand Upland

It covers most of the Hamirpur, Banda and the whole of the Jhansi districts. It is a broad undulating land ranging from 150 to 350 m in elevation and interspersed with small narrow and low ridges running parallel to the drainage lines in Jhansi district and tending to form clusters in Hamirpur and Banda districts.

The Mirzapur Upland

It can be divided into three sub-units: (i) The Vindhyan platueau in the north, drained by the Belan and the Karamnasa rivers, and bounded by the Kaimur plateau in the north and the Bijaigarh plateau in the south. The plateau surface is generally flat topped, with an average elevation of 300 m. It is broken by small hills and spurs here and there. (ii) The Son gorge is a narrow deep valley, generally 6 to 12 km wide lying between the precipitous scarp in the north and much lower and less difficult terrain of granitic and gneissic hill ranges in the Sonpar tract. (iii) The Sonpar upland consists of gorges, open basins, divides and dissected ridges. This region also exhibits a complex heterogeneous nature of the agents of erosion and the forces guiding their action. The valleys of the Bijul, the Rihand, the Kanhar and the Pandu as well as the divides in between them form micro scale landforms in this part. Parallel ridges, probably due to differential erosion, are observed on the divides in the Bijawars.

DRAINAGE

The chief drainage artery of the state is the Ganga, all other rivers drain to it directly or indirectly. The Ganga originates from the Gangotri in the Himalayas and flows from northeast to southwest up to Devaprayag after which it takes two sharp turns and flows from north to south from Rishikesh to Hardwar in its upper reaches. In the plain region it flows north to south in the western, northwest to southeast in the central and towards northeast in the eastern parts of the state. The main perennial tributaries of the Ganga are the Yamuna, the Ramganga, the Sai, the Gomti and the Ghaghra, which originate from

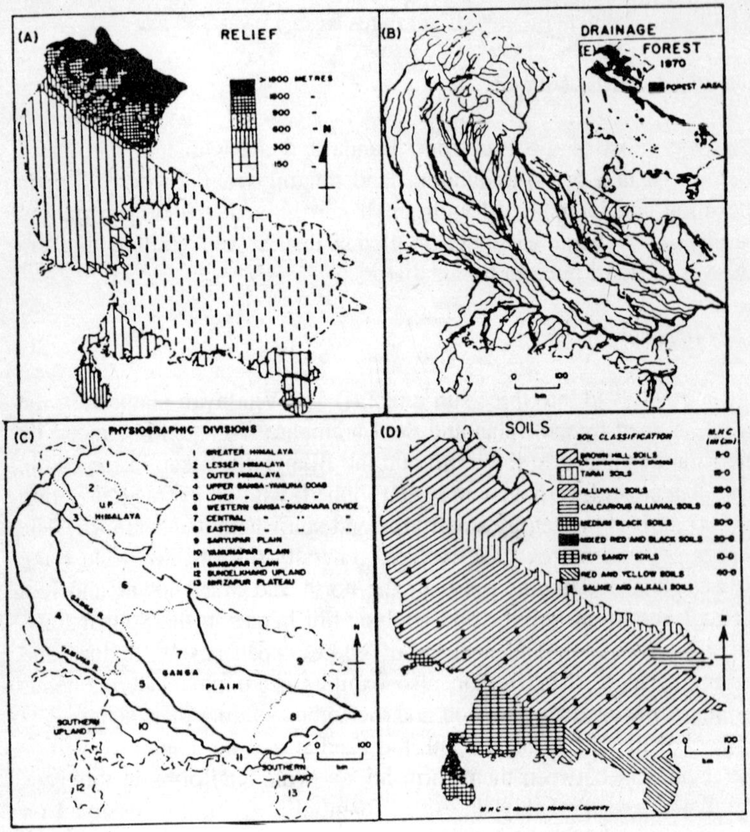

FIG. 1.2: (A) Relief, (B) Drainage, (C) Physiographic Divisions, (D) Soil, (E) Forest

the Himalaya with the only exception of the Gomti which rises from Pilibhit. The Yamuna river runs almost parallel to the Ganga till it joins the latter at Allahabad. The important tributaries of the Yamuna are the Chambal, the Betwa and the Ken all originating from the Deccan plateau. The Ramganga and Ghaghra also run almost parallel to the Ganga, the former joining it near Kannauj in Farrukhabad district and the latter near the eastern most tip of the state in Ballia district. The main tributaries of Ghaghra are the Sarda, Rapti and Burhi Gandak, all flowing from northwest to southeast. The Gomati is a river of plains and joins the Ganga beyond Varanasi. The river Gomati and

Ghaghra mostly inundate their flood plains during the rainy season and cause serious damage to human life, livestock and property. The Gandak, another important tributary of the Ganga, forms the northeastern boundary of the state for a short distance and then flows in Bihar to join the Ganga near Patna.

In the southeastern upland the Son is a major tributary of the Ganga, it runs parallel to the Ganga and joins it outside the state. The other tributaries of Ganga in this part are the Tons, the Karnauti, the Jirgo, the Ojhla, the Khajuri, the Chater Nali, and the Karmanasa, all emerging from the eastern Vindhyan upland and joining the Ganga as its right hand tributaries. The major tributaries of Son are the Bijul, the Rihand, the Kanhar and the Pandu, which flow from south to north over the Son-par upland.

The Himalayan rivers do more aggradation than those coming from the Vindhyan range and thus they contribute more in filling up the great plain with silt. They also provide immense potentiality for irrigation. They have perennial supply of water from monsoon rainfall supplemented by snow melt in the hot weather season. The Ganga, the Yamuna, and the Ghaghra have been used from historic period for long distance navigation, mainly by country boats.

SOIL

Soil formation is a very slow but ceaseless process. Soils *in situ* derive their chief characteristics to a great extent from the geological formation beneath them. The climatic and biotic agencies decompose the rock outcrops. These decomposed materials are sometimes left to rest over the rock, from which they are decomposed but are often transported to lower levels by agents of erosion. The importance of the present study of soil types and soil profiles lies in the fact that it enables us to make quantitative estimation of infiltration in the soil, water storage capacity, percolation, surface and subsurface runoff, tractionability of material and plant growth. Soil types of the state are classified into nine groups and for each group moisture holding capacity has been determined (Fig. 1.2D)

Brown-Hill Soils

Mukherjee and Das (1940) made a detailed survey of soils of the government orchard at Chaubitta.[4] Regarding the geology of the area, there are two main soil forming rocks-garnetiferous and biotitic schist and weathered granite and gneiss. Based on the pedogenic soil forming process under the climatic and topographic conditions, there are four types of soils, viz., red loam, brown forest, podsol and meadow soils. The red loam soils are found in dry places where the organic matter is rapidly mineralised owing to favourable a subsoil drainage conditions. The pre-eminent nature of profile in brown forest soil is the moderate organic status of the surface. The podsol soils are found under humid conditions. The meadow soils are usually found near *nallas* and cool shady continuously moist places.

Tarai and Bhabar Soils

Geologically these soils are subrecent alluvial formation. On the basis of topographical differences, they may be classified into three simple categories - (a) Soils of lowlying riverine tracts are silty or sandy, locally known as the '*bhat*' soils. They are highly calcareous and have alkaline reactions as well as good moisture holding capacity, (b) Soils of sloping surfaces are clayey loam, well drained and good for tillage. They are mildly calcareous and contain good amount of nitrogen, and (c) Soils of slightly elevated flat surfaces are clayey at surface, but in the lower layer show loamy texture resting on sandy subsoils. The surface clayey layers have high water holding capacity and are rich in calcium. They are neutral to alkaline in reactions. The nitrogen content ranges from 0.125 to 0.68 per cent.

Alluvial Soils

A detailed work on the alluvial soils of Uttar Pradesh has been reported by Mehrotra (1951, 1952, 1953 and 1954). The soils have shingle grained structure and textures vary from sandy to clayey.[5]

In the districts of Saharanpur, Muzaffarnagar and Meerut, the soils are generally heavy sandy-loam and of reddish-brown to yellowish colour. Further east, in the districts of Bareilly, Bijnor, Pilibhit and Moradabad, soils are generally heavy loam of light yellow colour. These soils are generally slightly alkaline, except in east Pilibhit where they are acidic.

In the northwestern parts of the districts of Kheri and Sitapur, soils are loam or sandy loam. In the eastern part of the Kheri and western part of Sitapur, the soil is of lighter texture and slightly alkaline. In the southeastern part of Sitapur, the soils are acidic in reaction. In the Hardoi district, the soils are sandy loam and slightly acidic in western part and slightly alkaline in the eastern part. The soils of Barabanki, Lucknow and Kanpur are slightly alkaline sandy-loam. In parts of Azamgarh and Jaunpur districts, the soils are sandy and slightly alkaline.

The eastern districts comprise two types of soils, *'bhat'* and *'bhangar'*. A third type of soil (*dhus*) also occurs near the river banks, which are liable to inundation. *Bhat* soils are found generally in low lying areas. They are sandy-loam and slightly alkaline to moderately alkaline in reaction. They possess clayey appearance and markedly high water holding capacity. *Bhangar* soils on the other hand, are generally found on higher surfaces and their texture varies from loam to sandy loam. The soils are usually slightly alkaline to slightly acidic in reaction.

Calcareous Alluvial Soils

These soils occur between the Ghaghra and the Tarai in the Deoria and Gorakhpur districts. The soils are loamy in texture and their clay content does not exceed 20 per cent. They are rich in lime and slightly acidic to slightly alkaline in reaction. Their nitrogen content is fairly above average and potash is well supplied.

Medium Black Soils

These soils are found in the southern parts of Agra and Jhansi districts. They have developed on crystalline rocks and are heavy textured. They are locally known as '*mar*' and *kabar*, the former being less clayey than the latter. They are calcareous, neutral to mildly alkaline in reaction and have higher capacity of retaining moisture.

Red-Sandy Soils

These soils have developed on crystalline rocks in the Bundelkhand and are locally known as '*rankar*' soils. They are *in situ* soil and are usually found on higher elevations. They are characterised by the presence of iron concretions in the profile. They also contain 30 to 50 per cent of stones.

Mixed Red and Black Soils

These soils are found in the northern part of Jhansi district. The black soils are deep, clayey in texture and occur in the low lying areas or flat landscape. The red soils are light textured varying from sandy loam to sandy clay on the uplands and their slopes.

Red and Yellow Soils

These soils occur in the southeastern part of the state mainly in the Mirzapur district. The common form of these red soils is sandy clay coloured by iron peroxide. Their colours vary from grey, brown, yellow of various shades to even very dark shades. They are rich in potash, have sufficient alkaline content and are generally deficient in nitrogen, phosphoric acid, humus and lime. These soils have developed on sedimentary Vindhyan formations and crystalline metamorphic rocks such as granites, gneisses, schists and other subordinate rocks. However, they are usually thin and shallow except in valley basin.

Saline and Alkaline Soils

These soils are found in patches over the alluvial tract. They vary from sandy-loam to loam. High content of alkali, excess of salts, paucity of nitrogen and plant nutrients are their characteristic feature. The salts, commonly found in the *Reh*, are sodium carbonate and sulphate together with calcium and magnesium compounds.

NATURAL VEGETATION

Many natural and biotic factors like climate, soil, drainage, man and animal, influence the distribution of natural vegetation. Hare (1954) and later Thornthwaite and Hare (1955) attempted to correlate potential evapotranspiration and the moisture index (I_m) with the distribution of vegetation.[6] Major in an alternative approach has considered a third factor, actual evapotranspiration (AE) and has suggested that vascular plant activity and growth might be related to the actual water loss of the vegetation since this factor expresses the real water activity of the plant rather than some unrealised potential value.[7] In the present study, the annual potential evapotranspiration values vary from 69.5 to 170.5 cm and moisture index 286.33 to 53.66

over the state. Large variations are found in the U.P. Himalaya due to orographic influence which directly affect the distribution of vegetation. The extent up to which the distribution of vegetation can be explained on the basis of climatic conditions depends mostly on the proper selection of climatic factors. The active factors of climate from vegetation viewpoint are water surplus or water deficit, actual evapotranspiration and soil moisture storage. The magnitude of these factors are helpful in determining the distribution of vegetation.

The forest land of the state comprises 39,770 km^2 representing 13.54 per cent of the total area. This is well below the all India average and far below the optimum (33.50 per cent), the desirable goal set by the National forest policy.[8] The proportion of area under forest in different districts of the state are above 40 per cent in the U.P. Himalayan districts and in Mirzapur, 15 to 40 per cent in Pilibhit and Kheri, 10 to 15 per cent in Saharanpur, Bijnor, Bahraich and Varanasi, 5 to 10 per cent in Gonda, Gorakhpur, Agra, Etawah, Jalaun, Hamirpur, Banda and Jhansi, and below 5 per cent in the rest of the Ganga valley districts.

Originally, the state had larger areas under forest which have gradually been cleared off for agriculture and settlement. Natural forests are found mostly in the U.P. Himalaya, its foothills and along a narrow belt in the plain, i.e., in the Tarai and Bhabar areas. Some forests also occur in the southern upland of the Vindhyan hills. Woodland are almost confined to the riverine strips in the Ganga valley. The grasses in the Ganga valley are Sarpat, Kans, Narkul etc. and bushes and trees are *Dhak (Buten monosperma)*, *Babul* (*Acacia arabica), Bamboo, 'Mahua' (Madhuca indica), Neem (Azadirachta indica), Ber (Zizyphus jujuba), 'Bargad', 'Peepal', 'Gular' (Ficus rasemosa), Bel (Aglemarmetos)* etc.

The chief forest types occurring in the state are:

1. Alpine scrub and low forests comprising shrubs like *Rhododendrom companulatum, Pyrus foliolosa, Funiperus recurua* etc., with grasses and herbs at elevations of around 3,600 m and sometimes may be seen in small patches even above.
2. Himalayan dry temperate forest occurs in the inner ranges of the Garhwal Himalaya (2,000-3,600 m). Main species are the conifers, *cadrus deodara, Abies webbiana, Abies pindrow, Picea smithiana* and *Pinus wallichana*, and the broad-leaved

associate species of *Acer* and *Fraximus.*

3. Himalayan moist temperate forest occurs between the subtropical pine and high level alpine zones. The main species are the conifers, Fir [*Abies pindrow* (1,200-2,000 m), *Abies webbiana*], Spruce (*Picea smithiana, Deodar (Cedrus deodara)* and Blue pine (*Pinus wallichiana*). Cypress (*Cirpressus torulosa*) are found on limestone area and broad leaved species of *Suercus* along the stream banks.
4. Himalayan subtropical forests (*Pinus roxburghu* or *chir*) occur at elevations ranging from 1,000 to 2,300 m.
5. Tropical moist deciduous forests are confined to the sub-mountain and Bhabar-Tarai areas. The main species are *Sal* (*Shores robusta*), *Sain (Terminalia tomentosa)* and *Dhawra* (*Lagerstroemia parfilora*), interspersed with moist mixed miscellaneous forests of *Haldu* (*Adina cordifolia*), *Simul* (*Salmalia malabarica*), *Kanju (Holoptelia integrifolia)*, *Gutul* (*Trewia nudiflora*), *Khair (Acacia catechu)* and *Sissoo* (*Dalbergia sissoo*).
6. Tropical dry deciduous forests are found in the dry places of the sub-mountain and Bhabar-Tarai areas. The main species are Sal. Other species are *Haldu, Khair, Sain, Tendu* (*Diospyros tomentosa*), Kardhai (*Anogeissus pendula*) and Bakli (*Anogeissus latifolia*).
7. Tropical thorn forests are found in the dry southwest region of the state, often as open scrub areas of *Babul* (*Acacia arabica*), *Acacia leucophlosa, Prospis spicigera, Balanites aegyptica* and *Salvadora oleoides.*
8. In the southern upland of Bundelkhand region *Dhak, Semal, Salai* and *Babul* are the main types of acacias. *Khair* is a common tree. *Hingota, Karonbha* and *Karrel* are mostly utilized for grazing. Grasses of various kinds such as *Musel, Uara, Gunea, Karta, Pasai, Dula, Kans, Daph* and *Gandar* etc. grow profusely in the rainy season. In the Mirzapur upland the common species among grasses are *Sawai* (*Ischoemun angustifolum*), *Kus* (*Posunosu ides*) etc. forming a dense under growth in the forest. Among the forest trees, *Amaltas, Harre,* and *Tendu* are of significance as they provide forest products for local commerce. In addition, *Khair* (*Acacia catechu*), *Palas* (*Butea frondosa*), *Mahua* (*Madhuca indica*), *Imli, Babul* (*Acacia arabica*) are also found on the upland.

It may be emphasized here that only the types three, four and five, described above, are of commercial importance for their timber, fuel wood or industrial wood. To a lesser extent, the types two and six also yield such produce. The other areas have to be maintained under forest cover mainly for soil conservation, ecological balance of the environment and other protective functions (flood control etc.), besides yielding some indirect benefits.

NOTES

1. Singh, R.L. (Ed.): *India: A Regional Geography*, National Geographical Society of India, Varanasi (1971), p. 445.
2. Wadia, D.N.: *Geology of India*, London (1953), p. 391.
3. Hari Narain: "Air Borne Magnetic Survey", Proceedings of Seminar on Earth Sciences, Pt. 1-Geophysics; Hyderabad the Indian Geophysical Union, (1965), pp. 119-129.
4. Mukherjee, B.K. and Das, N.K.: (i) Studies of Kumaon Hill Soils (ii) Effect of Terracing and Cultivation on Soil Types at Chaubitta, *Indian Journal of Agricultural Science*, Vol. 11 (1941), pp. 941-953.
5. Mehrotra, C.L.: *Soil Survey and Soil Work in U.P.*, Superintendent of Printing and Stationery U.P., Allahabad, Vol. I, (1951), Vol. II (1952), Vol. III (1953), Vol. IV (1954).
6. Hare, F.K.: "The Boreal Conifer Zone", *Geographical Studies*, Vol. 1, London (1954), pp. 4-18; Thornthwaite, C.W. and Hare, F.K.: "Climatic Classification in Forestry", *Unasylva*, Vol. 9, No. 2, Rome (1955), pp. 50-59.
7. Major, J.: "A Climatic Index to Vascular Plant Activity", *Ecology*, Vol. 44, Indiana (1963), pp. 485-98.
8. NCARE: *Techno-Economic Survey of India*, U.P., New Delhi (1965), p. 68.

2

SPATIO-TEMPORAL VARIATIONS IN CLIMATIC ELEMENTS

IT IS necessary to know weather and climate. The weather or climate of any place is the sum of its atmospheric conditions, e.g., temperature, pressure and wind, precipitation, humidity etc. Therefore, the elements of weather and climate are the same, but weather is a state of the atmospheric conditions at a given time and space, whereas climate is a composite picture or generalization of the variety of day-to-day weather conditions.[1]

It is, therefore, obvious that weather varies from day-to-day and climate differs from place to place depending upon the variations in amount, intensity and areal distribution of their elements. In the present study region of Uttar Pradesh, these elements differ widely over the mountain ranges and the valleys, plain and plateau, because of (i) latitude causing differential effective insolation, (ii) altitude, (iii) mountain barriers, (iv) air masses and their movement, (v) extension of the semi-permanent (seasonal) high and low pressure areas in the form of wedge and trough, and (vi) storms of various kinds and their frequency. These factors cause changes in the climatic elements particularly in rainfall and temperature, which lead to varieties of weather and climate.

Before discussing and analysing the variability in climatic elements, a brief description about the controlling factors becomes necessary. Out of these the aspect of mountain barriers and their altitude influence the intensity of insolation because of variation in the angle of incidence of the sun's rays on the sloping surface. Latitude is not a very significant factor because of its short extent. The inclination of sun's rays controls the solar radiation that the surface receives during the summer, winter and the whole year which is dependent upon season, latitude, degree of slope and the direction towards which it faces. Altitude and mountain barriers govern the uneven distribution of rain and temperature in the valleys and ranges of U.P. Himalaya. It may be compared with the uneven distribution of rain and temperature over the plain and southern uplands of the state. The latter three controlling factors are uncertain from year to year and time to time and hence make variable weather conditions from time to time and place to place. Effects of these factors become evident while analysing the climatic elements individually for a clear understanding of climate.

TEMPERATURE

Temperature is an independent variable among the climatic elements. It causes change in the pressure distribution and consequently in wind direction and its velocity which control atmospheric humidity, condensation, formation of cloud and their drafting in the sky, precipitation and storms. So a detailed study and analysis of temperature condition over the state, its areal and seasonal distribution, annual variation of its magnitude and distribution of temperature anomalies are necessary, because temperature, along with its controls over other elements of climate, determines the character of local climate.

In the mountains, the temperature decreases by approximately 0.5 to 1.0°C with 100 m² (average 0.6°C/100 m) increase in height. Therefore, all temperatures have been reduced to sea level to draw isotherms. It is also affected by the nature of the surface. Topographic features like hills and trends of the hilly ranges, the presence of lakes, vegetation or desert largely determine the variation in magnitude of temperature. Thus irregularities of terrain account not only for horizontal variations in the air temperature but also cause formation of frosty hollows and warm patches in parts of the U.P. Himalaya.

Horizontal Distribution of Temperature and its Variations

Temperature distribution over the state in January and July are shown by means of isotherms. January is the coldest month while July isotherms present a typical distribution of temperature as the southwest monsoon bursts and modifies the maximum temperature in the month of June. The error in the reduced values is not likely to exceed 1°C for the stations with altitude of less than 1000 m³, but in the U.P. Himalaya at stations like Mussoorie and Mukteswar which are more than a kilometre a.s.l., the error is likely to exceed 1°C.

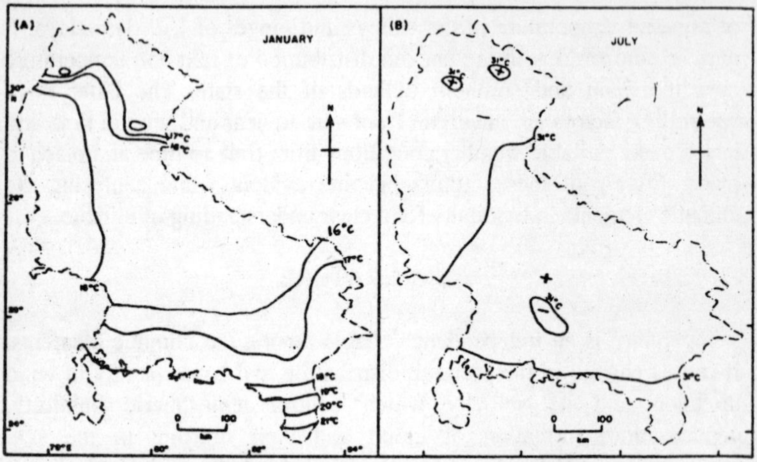

FIG. 2.1: Isotherms: (A) January, (B) July

(a) Figure 2.1A shows the mean isotherms for the month of January, which vary from 16°C to 21°C. The temperature increases towards the south to above 21°C. Higher temperature in the Ganga plain with a comparatively lower values in the west is the result of continentality and effects of the western disturbances. Highest values are observed over the Mirzapur and Bundelkhand regions recorded at Pipri and Jhansi. Compared to the plain temperature is higher by 1°C to 2°C at the U.P. Himalayan stations, because of error in reduced values.

(b) Figure 2.1B shows the mean isotherms for the month of July. With the spreading of the maritime airmass of the southwest monsoon over the state, temperature is almost uniform except over the U.P. Himalaya. The temperature is a little higher in the west than in the east due to continentality and lesser cloudiness. Temperature also increases from north to south, the isotherm of 31°C passes along the northern fringe of the southern upland.

(c) The mean annual temperature is shown in Fig. 2.2A. The distribution of mean annual temperature decreases from south to north. Comparatively higher temperature is observed at Pipri and Jhansi. The gradient of temperature is higher in the south than in the north. The mean annual temperature varies from 26° to 30°C over the state. Compared to the plain the mean annual temperatures are also higher (about 1°C) in the U.P. Himalaya because of consideration of reduced values (at normal rate) which are prone to errors.

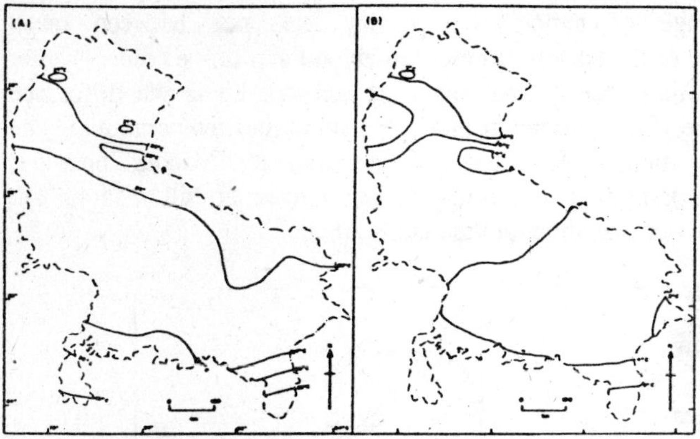

FIG. 2.2: (A) Mean Annual Temperature (Reduced to Sea Level), (B) Temperature Anomaly (Mean Annual)

Temperature Anomaly

Temperature figures for different latitudes depending upon the amount of insolation on an imaginary homogeneous surface of the earth is available. If this temperature be compared with the temperature of any place (reduced to sea level) the effects of altitude is eliminated. Any difference between these two values will indicate the effects of other factors like distance from the sea, nature of terrain etc. This difference is called temperature anomaly. The line joining places of equal temperature anomaly is called an isonomalous line.

Mean annual temperature anomalies for different stations have been shown in Fig. 2.2B. It is greater in the U.P. Himalaya, because cold polar air encroaches in winter and orographic and convectional lift of southwest monsoon causes heavy rainfall and consequent release of latent heat in summer. Appreciable difference in mean annual temperature (reduced to sea level) between the plain and southern upland results in greater anomalies on the latter. In the plain, temperature anomalies are higher in the west than in the east because the western parts are highly affected by cold air in winter which produces large anomalies.

Range of Temperature

Annual and diurnal ranges of temperatures are shown in Fig. 2.3A. The annual range of temperature is the difference between mean temperature of the hottest summer month and that of the coldest winter month; similarly the diurnal range of temperature is the difference between the mean maximum and mean minimum temperatures. 'The ranges are mainly functions of continentality, vapour pressure, cloudiness, latitude and altitude'.[4] The ranges at hill stations are generally lower than those at the plain stations.

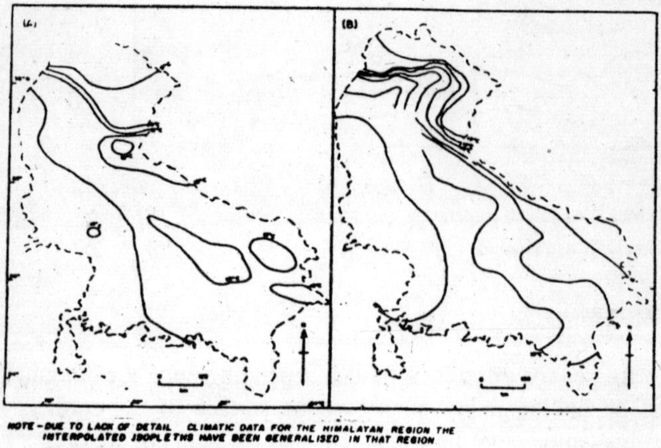

FIG. 2.3: (A) Mean Diurnal Range of Temperature, (B) Annual Range of Temperature

(a) The diurnal range of temperature varies from 8.1° to 14.1°C. It is least in the U.P. Himalaya varying between 8.0° and 12.0°C (except at Tehri 14.0°C) and increases to as much as 14.1°C towards the southwest. In the rest of the state, it is roughly 13°C.

(b) The annual range of temperature varies from 12.3° to 20.2°C. The gradient of annual range of temperature is higher in U.P. Himalaya in comparison to that in the plain. The isopleths are almost parallel to each other with values increasing towards the southwest. It is quite high at Agra (20.2°C) and low at Mukteswar (12.3°C).

Other Significant Features

(a) Extreme range of temperature is the difference between the highest and the lowest temperature recorded over a long span of time. The highest temperature is reached in the months of May or June and the lowest in the month of January or February. The highest and lowest temperatures recorded during an eighty-year period at selected stations are shown in Table 2.1.

(b) The diurnal temperature variation is considerably modified by the maritime airmasses causing increased cloudiness, which obstructs the sun's rays. Thus, day temperature decreases instead of increasing.

(c) The highest temperatures are in June after which the effect of the southwest monsoon brings down the temperature.

(d) Temperature rises again towards the end of the monsoon and there is a second maximum in September. Decrease of cloudiness mainly accounts for this second rise in temperature.

TABLE 2.1: Highest and Lowest Temperature During an Eighty-Year Period

Stations	Highest Temperature in °C		Lowest Temperature in °C	
1. Mussoorie	34.4	May 24, 1949	−6.7	Feb. 10, 1950
2. Mukteswar	30.6	July 1, 1901	−7.8	Feb. 10, 1905
3. Dehra Dun	43.9	June 4, 1902	−1.1	Jan. 11, 1944 Feb. 1, 1905
4. Roorkee	46.7	June 13, 1932	−2.2	Feb. 2, 1905
5. Bareilly	46.7	May 29, 1884	0.0	Feb. 2, 1905
6. Aligarh	46.3	June 17, 1958	0.6	Jan. 16, 1935
7. Agra	48.3	June 2, 1889	−2.2	Jan. 16, 1935
8. Mainpuri	47.8	June 8, 1948	−1.7	Jan. 11, 1927
9. Jhansi	47.8	June 1, 1924	0.6	Feb. 2, 1929
10. Kanpur	47.2	May 30, 1941 June 11, 1931	0.6	Feb. 2, 1905
11. Lucknow	48.3	June 12, 1901	1.1	Jan. 3, 1946
12. Bahraich	46.7	June 6, 1958	0.6	Jan. 15, 1936
13. Gonda	49.9	May 8, 1958	1.1	Jan. 15, 1933
14. Gorakhpur	48.3	May 26, 1958	1.7	Jan. 15, 1933
15. Fatehpur	47.2	May 26, 1952	1.1	Jan. 20, 1935
16. Allahabad	47.8	June 12, 1901	1.1	Feb. 2, 1905
17. Varanasi	47.2	June 12, 1901	1.7	Feb. 2, 1905

PRESSURE AND WINDS

The distribution of sea level pressure over the state undergoes a complete reversal from January to July, so that a description of mean annual pressure pattern has no significance. January, which is the month of northeast trade has a low pressure gradient towards the south. The pressure gradient is towards the north in July during the southwest monsoon. Figures 2.4A and 2.4B[5] show the mean isobar and winds in January and July over the state, but it is not true that winds always blow from northwest in January and from southeast in July. Figures 2.4C, 2.4D and 2.4E indicate the directional pressure gradient responsible for particular types of wind patterns in winter and summer and their adjustment with the physical environment. In January the state is affected by western disturbances, which move towards NE and ENE. Consequently, they displace the high pressure area in the northwest and change the direction of winds, which is apparent from Fig. 2.4C. In July, with the development of a low over Uttar Pradesh, Madhya Pradesh and Bihar and a seasonal trough running from northwest to southeast, winds blow towards west and northwest in the western part of the trough and towards south to southeast in the eastern part of the trough Fig. 2.4D. When the low pressure shifts towards the west over the west U.P. and adjoining Rājasthan, all isobars trend towards the low and wind blows towards the northeast which is quite apparent from map Fig. 2.4E.

RELATIVE HUMIDITY

Annual relative humidity has been depicted by shading whereas monthly mean relative humidity at 0830 and 1730 hrs have been shown by graphs for 13 representative meteorological stations in the state (Fig. 2.5).

Annual relative humidity varies from 70 to 80 per cent in the U.P. Himalaya except in the river valleys and the leeward side of the ranges. It varies from 60 to 70 per cent in the Tarai, east U.P. and Mirzapur district. Lowest relative humidity (below 30 per cent) is confined to Agra and Mathura districts. In general, humidity increases as one moves from the southwest towards the east or north.

The pattern of monthly relative humidity remains almost similar at all the stations. Two maxima, one in winter and the other in rainy season, and two minima, in spring and autumn, are characteristics of all curves showing monthly relative humidity for different stations.

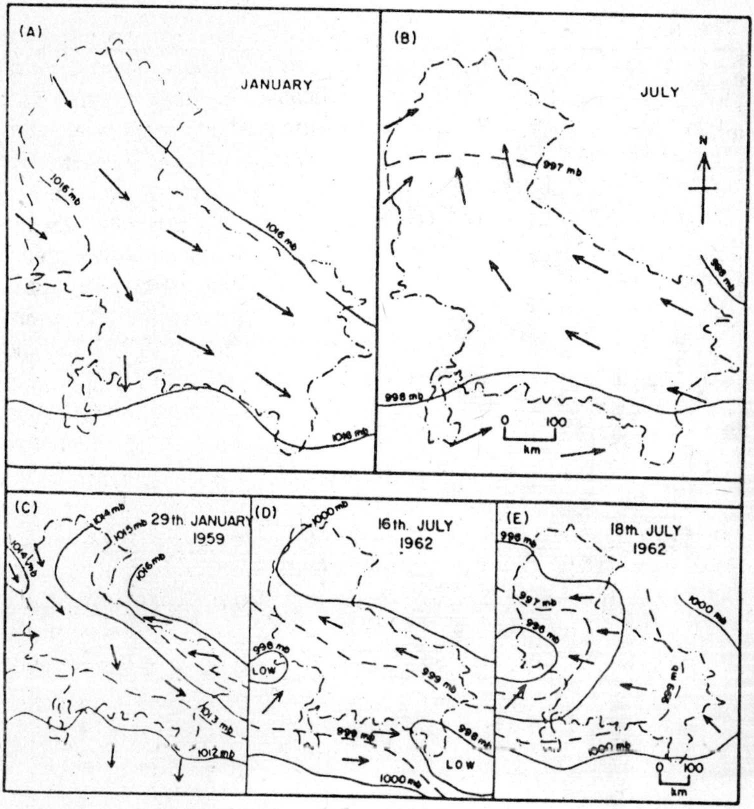

FIG. 2.4: Pressure Distribution and Surface Wind; (A) January, (B) July, (C) Disturbed by the Western Disturbances, (D) and (E) Disturbed by the Bay Depressions

The relative humidity varies inversely with temperature during the day. The atmosphere is most damp in the morning and dryness increases as the temperature rises during the day time. It increases again in the evening hours as the temperature falls. The relative humidity curves for the morning (0830 hrs) and evening (1730 hrs) run almost parallel to each other. It in higher is morning than that in the evening over the plains throughout the year. On some hill slopes it is higher in the evening than in the morning except in the dry summer months of May and June it may be due to inversion of temperature.

The diurnal variation in relative humidity is greater in the winter than in the summer months.

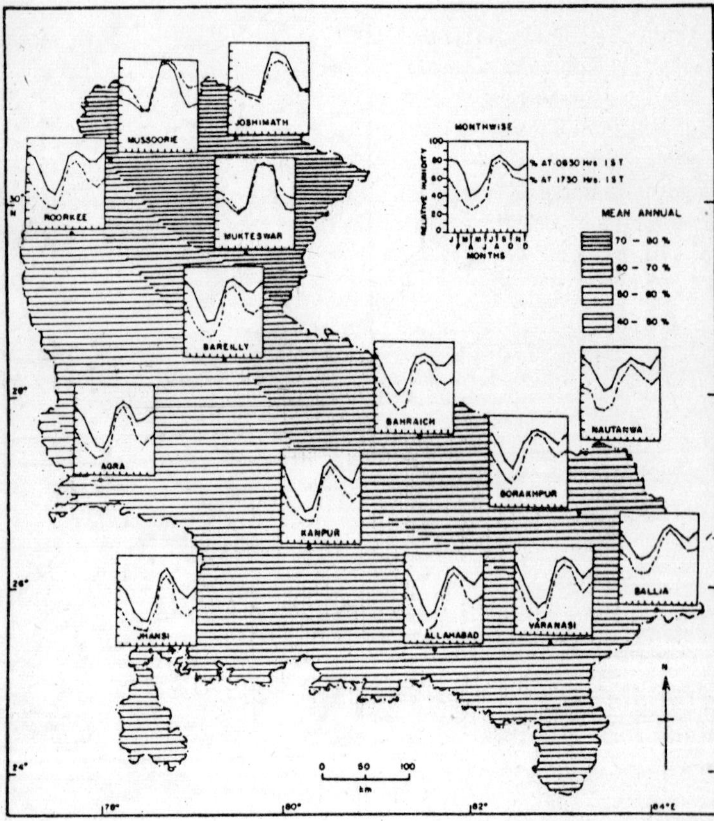

FIG. 2.5: Relative Humidity

RAINFALL

Mean Annual Rainfall

The rainfall data for 200 rain gauge stations including the meteorological stations have been analysed for a period of 50 years (1901-50). The mean isohyets are shown in Fig. 2.6A. Striking variations in rainfall amount are in the U.P. Himalaya. There occurs heavy orographic rainfall on the windward side of the ranges with a rapid decrease of rainfall on the leeward side (160 cm to above 200 cm on the windward side and 80 to 160 cm on the leeward side). Secondly, the amount of precipitation also increases with elevation (Mussoorie 269 cm and Nainital 254 cm) except over the northern snowclad areas. The third striking variation is between the river valleys and the hill stations, it varies from 98 to 250 cm. In the mountainous areas rain recording stations are few, particularly at higher elevations and their is additional difficulty in measuring the amount of snowfall. Rainfalls measured at stations situated in the river valleys are not representative of the mountain crests and their slopes. In the plain, rainfall decreases from east to west. The lowest value is about 70 cm at Agra from where it increases towards the foothills of the Himalaya and eastern U.P. In the Tarai tract rainfall increases probably due to the influence of the Himalaya. Even in this tract the rain varies from 120 cm in the east to 160 cm in the west, but in the middle part it is below 120 cm.

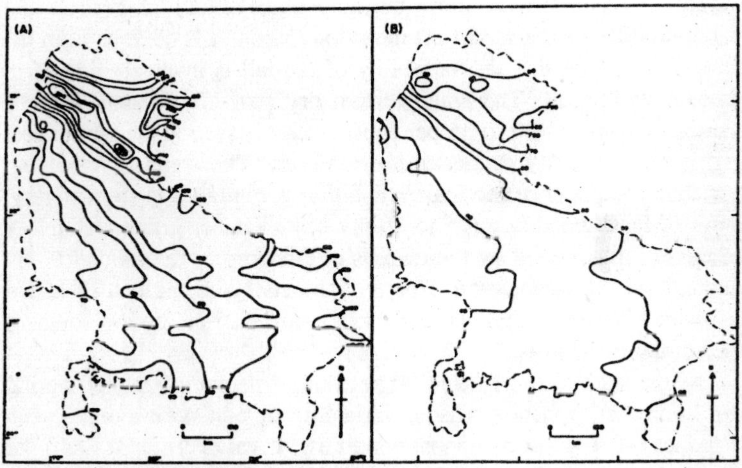

FIG. 2.6: (A) Average Annual Rainfall, (B) Average Annual Rainy Days

Rainfall is relatively higher over the southern uplands of Mirzapur and Bundelkhand than over the plain, the amount being above 100 cm. It is probably due to the fact that these parts are affected by the western disturbances in winter and by monsoon currents from the Arabian sea in addition to those from the Bay of Bengal in summer.

Rainfall Variability

It is experienced that rainfall varies considerably both in space and time from year to year. Some years stand out as phenomenal years, when the rainfall is either very much in excess or is much in deficit. Coefficients of variability (CV) have been calculated as:

$$CV = \frac{S}{M} \times 100$$

Where, *S* is standard deviation
M is the mean of the annual rainfall for 50 years.

In the case of variability in the southwest monsoon season (June to September), *M* represents the mean of the southwest monsoon rainfall. Figures 2.7A and 2.7B show the annual variability of rainfall and variability during the southwest monsoon season. It is obvious from the map (Fig. 2.7A)[6] that the variability of rainfall is inversely related to amount of rainfall. The southwestern dry part of the state has the highest variability (30 to 35 per cent) whereas towards the rainier east and north variability decreases progressively. The areas of maximum variability experience the least reliability of rainfall. On the contrary, areas of least variability (15 to 20 per cent) witness greater reliability of rainfall as revealed by the records of Gorakhpur, Deoria and Ballia districts on the northeastern fringe of the state and the belt of heavy rainfall in Nainital, Garhwal and Dehra Dun districts on the southern face of the Himalayas.

As the state receives 70 to 90 per cent of its annual rainfall during the southwest monsoon season, variability of rain during this season indicates the trend of its annual variability. It varies from 20 to 40 per cent in the state. Figure 2.7B justifies the aforesaid statement to a considerable extent. The variability of rain in the month of July is lower during this season, because the monsoon is then at its active stage. The important aspects of variation in rainfall during the southwest monsoon season are:

(a) The time of commencement of the southwest monsoon over the state.

(b) The pattern of distribution of rainfall including the intermittent breaks in monsoon.
(c) The time of withdrawal of monsoon.
(d) The variation in the total amount of rainfall during the season.

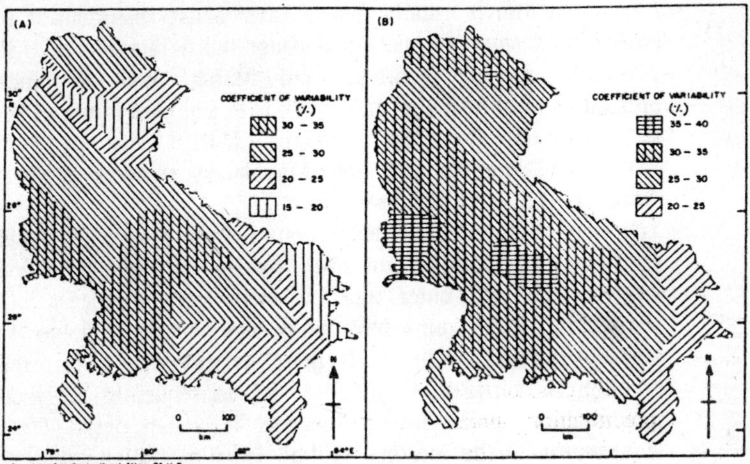

A - based on Agricultural Atlas Of U.P.

FIG. 2.7: Variability of Rainfall: (A) Annual Variability, (B) Variability in Monsoon Season

The variability is also high during other months indicating the highly erratic nature of rainfall during these months. Therefore, other causes in temporal and spatial variations of rainfall are atmospheric disturbances, i.e., thunderstorms, western disturbances and tropical storms and depressions.[7]

(a) The rainfall pattern follows closely the arrangement of thunderstorm cells and reflects to a considerable extent the stage of development of the different cells. Duration of rain at any place depends upon the number, size and rate of movement of these cells. In winter season thunderstorm rain is least. In the hot weather season thundery activity increases gradually, and during the southwest monsoon season it reaches its maximum stage. It is generally observed along and near the monsoon trough. As the monsoon trough shifts southward by October the thunderstorm activity also shifts with it. In the post-monsoon season, heavy rainfall roughly occurs in the areas of thunderstorm rain.

(b) The rainfall by western disturbances is usually from November to April and sometimes in the month of May. The amount of rain during these months is not uniform over the region. From January to March usually two or three active disturbances in each month enter the state and cause much rain in the U.P. Himalaya. During February and March there is much precipitation in the plains of western U.P. and their minor effect is observed in the eastern U.P. Generally, in November-December and April-May heavy showers of short duration occur with thunderstorms.

(c) The tropical storms and depressions largely govern the local rainfall. Such storms form over the Bay of Bengal or the Arabian sea, and enter into the state from southeast and southwest. Heavy rains have been experienced with slowly moving tropical cyclones. It these systems move fast the duration is considerably reduced and consequently the total precipitation under their influence is much less. Their contribution to the seasonal rainfall is of maximum intensity during the southwest monsoon. The intensity associated with these storms does not exceed 20 mm per day in the hot weather season and 30 mm per day during the post-monsoon season. In the winter season the rainfall intensity associated with these cyclonic disturbances is quite low and ranges from 5 to 10 mm a day.

Number of Rainy Days and Cloudiness

Figure 2.6B depicts the distribution of the average annual number of rainy days. The pattern of rainy days is similar to that of annual rainfall. Areas having more than 70 rainy days lie in the U.P. Himalaya excepting its leeward side. Number of rainy days varies between 50 and 70 in the Tarai and east U.P. Less than 40 rainy days are confined in the southwestern parts of the state. In the plain the number increases from west to east and towards the Himalayas. They are also higher in comparison to plain over the Mirzapur upland (60 days) (Appendix G).

The four selected months July, November, January and May are the seasonal representatives of cloudiness condition in a year, while the rainiest month happens to be July when more than three-fourths of the sky is covered with cloud (usually overcast). Clouds cover gradually decreases in the subsequent months and less than one-eighth of the sky remains covered with clouds in November. Almost one-fourth of the sky remains covered in January because of the western disturbances which bring much rain in the winter months. Cloud cover again decreases gradually allowing bright sunshine for most of the days when clouds cover is less than one-eighth during the summer months.

FREQUENCY OF OTHER WEATHER PHENOMENA

The state is also affected by fog hail, thunder, duststorm and squall. The average number of days in a year with these phenomena is shown in Table 2.2 for different meteorological stations. Hailstorms occur from November to May in the U.P. Himalaya and from January to April in Western U.P. as well as in the Tarai. On rare occasions, it also occurs in east U.P. in the months of February, March as well as in April. However, it is a rare phenomenon depending much upon local condition. Strong convectional current occurring anywhere starts sucking up moisture laiden winds or clouds to exceptionally high altitude causing condensation in large quantity below the freezing point and leading to the formations of hailstorms. Thunder activity is generally reported in all months, but it is dominant from June to September (the southwest monsoon season). Its frequency is high in the U.P. Himalaya, the Tarai and southwest as well as at Allahabad and Varanasi. Fog occurs from January to March. Its frequency too is high in the hills as compared to the plain. Duststorm and squall are dominant features during the hot weather season. They occur over the whole state except in major parts of the U.P. Himalaya.

CLIMOGRAPHS AND HYTHERGRAPHS

Climographs help in assessing weather condition with respect to human comfort in different months whereas hythergraphs indicate the suitability of climatic condition for different crops in a region. It is for this point of view that climographs and hythergraphs for 12 selected stations of the state have been drawn.

1. The climographs show interrelationship between two variables, the monthly averages of wet bulb temperature and relative humidity. Twelve points corresponding to each month are plotted on the frame. By joining these points a twelve-sided polygon is obtained, which is called a climography. Taylor has also suggested technical terms like

raw, muggy, scorching and keen for extreme cases denoting cold and damp, hot and damp, hot and dry, cool and dry weather respectively. Limits of these values have been indicated in the key of Fig. 2.8. The graphs approaching these four corner areas indicate uncomfortable weather conditions. 'Scorching' weather prevails in the months of April, May and first half of June over the whole state excepting in the U.P. Himalaya at the representative stations, i.e., Mussoorie, Mukteswar and Joshimath. Muggy condition prevails in the months of July, August and September.

The climographs at a glance reveal the fact that comfortable weather conditions prevail in the months of March, April, October and November over the plain area as well as over the southern plateau, whereas the months of May and June provide quite comfortable living conditions over the ranges of the U.P. Himalaya.

TABLE 2.2: Average Annual Frequency

Stations	Hail	Thunder	Fog	Dust-storm	Squall
1. Mussoorie	9.0	60	67.0	0.3	3.0
2. Mukteswar	3.0	54	62.0	0.0	0.0
3. Dehra Dun	1.6	70	3.0	0.3	0.0
4. Roorkee	2.0	39	1.5	3.0	0.0
5. Bareilly	1.7	31	15.0	4.0	0.5
6. Aligarh	0.7	32	1.9	5.0	0.0
7. Agra	0.5	16	5.0	5.0	4.0
8. Mainpuri	0.3	18	1.4	3.0	0.9
9. Jhansi	0.2	13	1.1	0.3	0.0
10. Kanpur	1.2	21	5.0	5.0	0.0
11. Lucknow	0.5	18	6.0	2.0	0.2
12. Bahraich	0.4	22	1.1	1.7	0.0
13. Gonda	0.8	18	1.3	1.3	0.3
14. Gorakhpur	0.0	6	6.0	1.1	0.0
15. Fatehpur	0.5	21	1.1	3.0	0.0
16. Allahabad	0.4	57	4.0	8.0	5.0
17. Varanasi	0.3	31	5.0	1.1	0.1

2. Like climograph, hythergraph is also a twelve sided polygon showing average monthly temperature and rainfall. Hythergraphs have been drawn for the 12 stations located in different parts of the state (Fig. 2.9). These are helpful in assessing the suitability of the climatic condition in different parts of the region for different crops. The

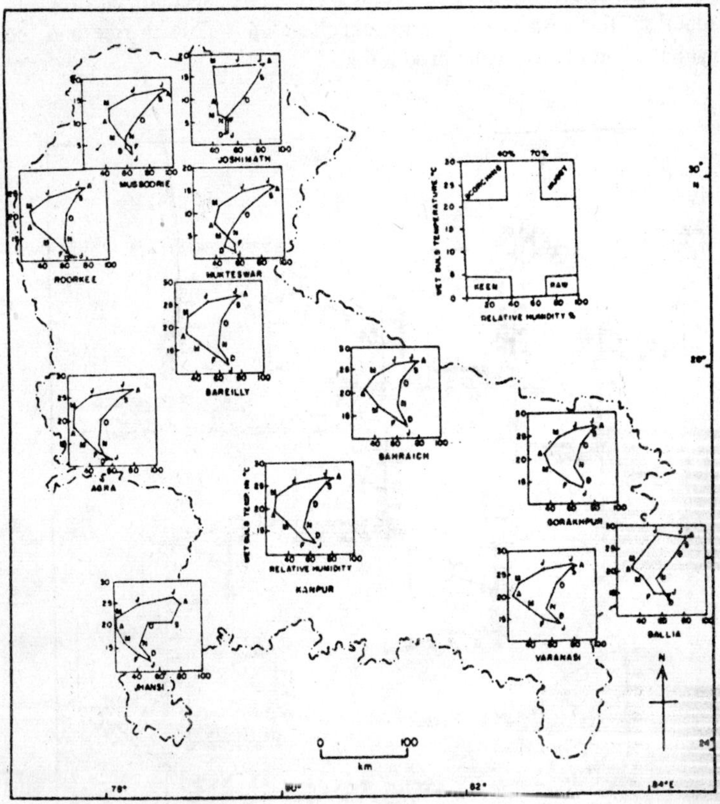

FIG. 2.8: Climographs

dominant crops are wheat, rice and sugarcane in the state. If these be compared with the ideal hythergraphs of Havana (Cuba) for sugarcane, of Hong-Kong (China) for rice and of Buenos Aires (Argentina) for wheat, they will be helpful in assessing as to how much differences with respect to rainfall and temperature exist in different parts of the state for good production of these selected crops. It may be pointed out that wheat, rice and sugarcane are produced in the state on quite a large scale although the amount of rainfall apparently appears to fall short of the ideal requirements for these crops. This deficiency is, however, compensated by increasing development of gravity and lift irrigation in different parts. The western and central parts have specialized in wheat production, the eastern most part and the Dun Valley in rice and the Saryupar plain in sugarcane production.

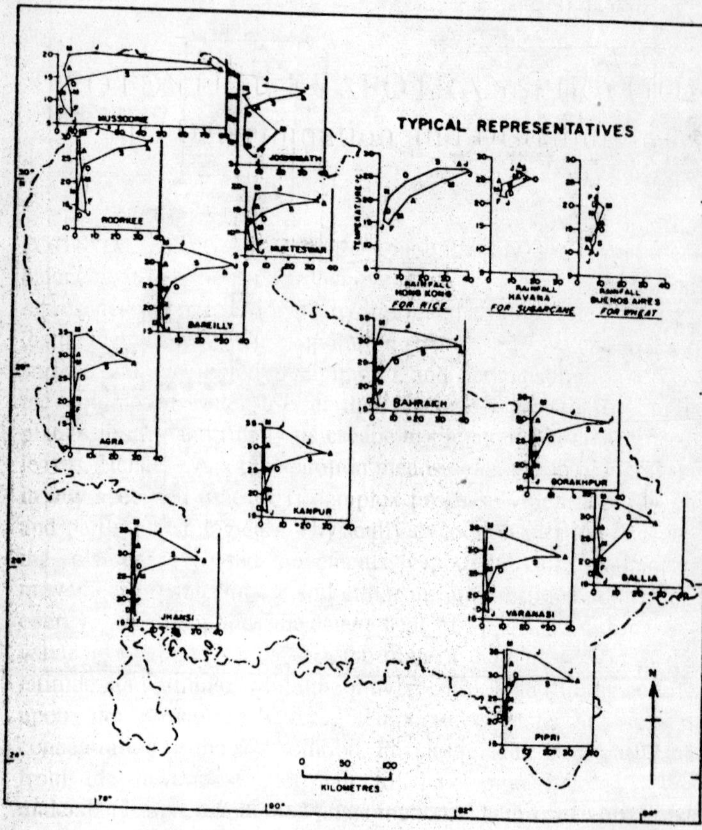

FIG. 2.9: Hythergraphs

NOTES

1. Trewartha, G.T.: *An Introduction to Climate*, McGraw-Hill Book Company, New York (1968), p. 2.
2. Rusin, N. Flit L. : *Man Versus Climate*, Moscow, p.44.
3. Rao, Y.P. and Ramamurthy, K.S.: Climatology of India and Neighbourhood, *Climate of India*, I.M.D. (1968), p. 4.

4. Rao, Y.P. and Ramamurthy, K.S.: op. cit., p. 6.
5. Rao, Y.P. and Ramamurthy, K.S.: op. cit. [Mean pressure (mb) and surface wind, Figs. 2.1 and 2.3].
6. Tewari, S.B.: *Agricultural Atlas of Uttar Pradesh*, Plate No. 9, (1971).
7. Jagannathan, P.: Rainfall in India, *Forecasting Manual*, Part IV, 13 Aug. I.M.D. (1968).

3

SEASONS AND THEIR CHARACTERISTIC FEATURES

THE YEAR in this monsoon country is popularly divided into three seasons, the cold season from mid-October to mid-March, the hot season from mid-March to mid-June and the rainy season from mid-June to mid-October. But the India Meteorological Department has recognised four seasons. The period from March to May has been designated as hot weather season because of high temperature, June to September as the season of the southwest monsoon, October to December as post-monsoon season and January-February as winter (cold weather) season. It is quite a convenient division for the weather study of our country. Since the state of Uttar Pradesh is situated in the northern part of the country, the southwest monsoon over here every year retreats completely by the end of November and the cold weather commences from December, hence December has been included in winter season in this study.

With respect to wind system, the state experiences winds from northwest, west and southwest in the winter season. The state is also affected in this season by the 'western disturbances'. These disturbances in the form of series of cyclones originating over the Medeterranian and sometimes even over the Atlantic Ocean are carried to our country owing to the southward shift of the wind system in this season. Sudden appearance of a feeble low pressure trough penetrating into the regional high pressure area, marks the advent of western disturbances. If two or more closed isobars appear in this trough it is designated as a depression. The disturbances revive when they take a more southerly trajectory and fresh moist tropical air is introduced into it resulting in the formation of secondary depressions or disturbances. 'The secondary lows are formed in the gulf of Aden as well as off Mekran, Kadhiawar and Konkan coasts, the last two of which go by the name of 'Gujarat and Deccan lows'. They move almost ENE and affect northern and central regions of the state.'[1] Precipitation and thunderstorms usually occur due to the combined influence of these secondaries.

During the summer, the wind is southwesterly from the Arabian sea but more or less easterly and southeasterly from the Bay of Bengal although its direction is controlled differently by orographic features. It advances from the Arabian sea and generally meets the Bay of Bengal current along the axis of the low pressure trough which normally extends down the Ganga Valley from northwest India to Orissa. This trough governs the wind direction in this state which is usually northeasterly, easterly and southwesterly in summer.

THE SOUTHWEST MONSOON SEASON
(June to September)

Temperature

It is clear from Fig. 3.1A that in July the isopleth of mean daily temperature of 28°C runs parallel and close to the foot of the Himalayas and that of 30°C is confined to the plains of west U.P. because, the southwest monsoon sets in early over east U.P., which mitigates the temperature in this region. The trend of the mean maximum temperature is similar to that of the mean daily temperature. The isopleths of the 32°C and 34°C are confined to the foot of the U.P. Himalaya and the west U.P. plain respectively. The isopleth of mean minimum temperature of 24°C almost coincides with the mean maximum temperature of 32°C, but the isopleth of 26°C runs right through the plains of west and east U.P. (Fig. 3.2A). The isopleth, however, also appears in the southern parts near the northern edge of the southern uplands.

Wind

The main factors affecting the wind direction and speed are the periodic oscillation of the axis of the trough of low pressure as well as invasion of a series of feeble or strong depressions into or close to this region. When the seasonal trough is well marked and is to the south of its normal position or when a deep depression moves across Madhya Pradesh, the winds over the state become strong easterlies which attain velocity of 40-50 knots. When the axis of the trough shifts to the foot of the Himalayas, strong westerly to southwesterly winds (40-50 knots) prevail. In general the surface winds alternate from easterly to westerly (through southerly with northward shift of the axis of trough and through northerly with southward shift of the axis) during the season and their speed averages to be 5-10 km/hr over east U.P., 3-7 km/hr over west U.P. and 3-5 km/hr over the U.P. Himalaya (except Mukteswar where it averages above 10 km/hr).

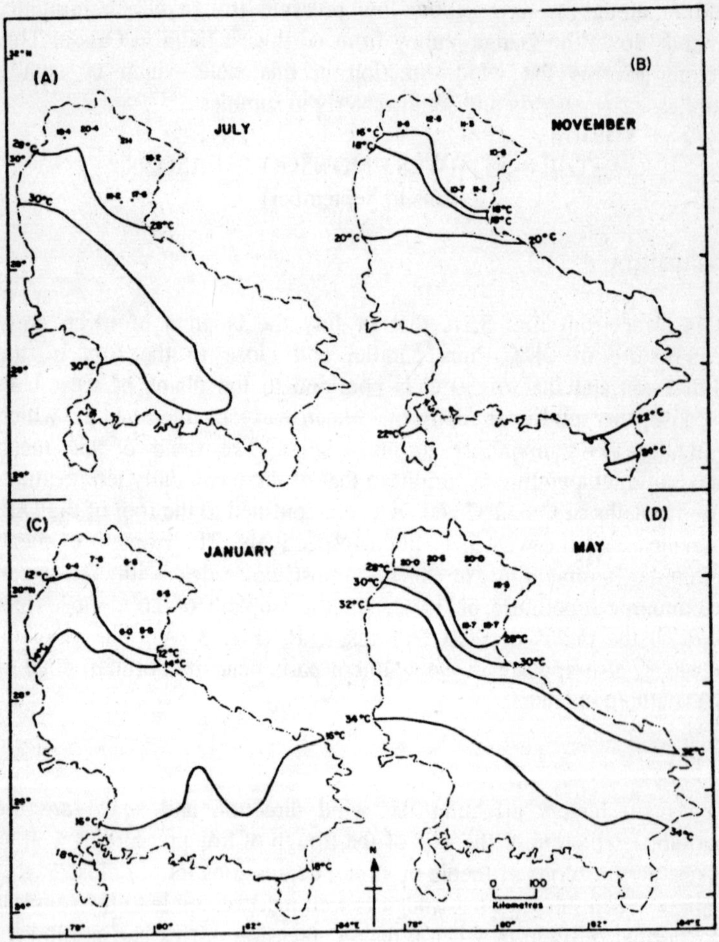

FIG. 3.1: Mean Monthly Temperature (Actual): (A) July, (B) November, (C) January, (D) May

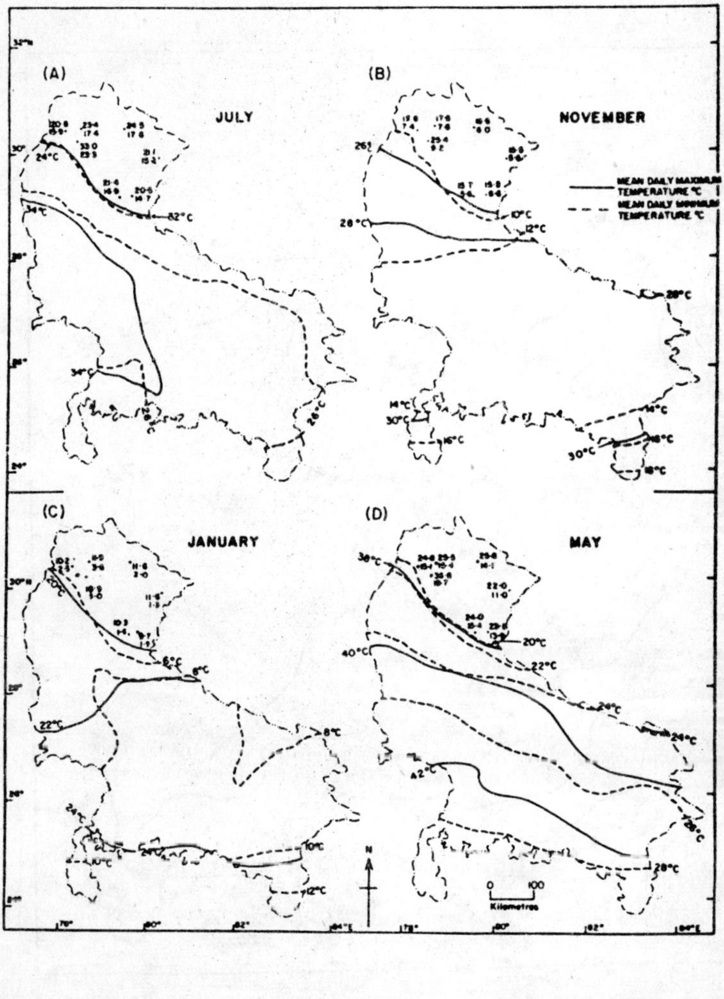

FIG. 3.2: Seasonal Distribution of Actual Temperature (Mean Daily Maximum and Mean Daily Minimum): (A) July, (B) November, (C) January, (D) May

Relative Humidity

Highest percentage of relative humidity occurs during this season which varies from 75 to 90 per cent over the U.P. Himalaya and 65 to 75 per cent over the rest of the state except Agra and Jhansi districts where it is below 65 per cent (Fig. 3.3A).

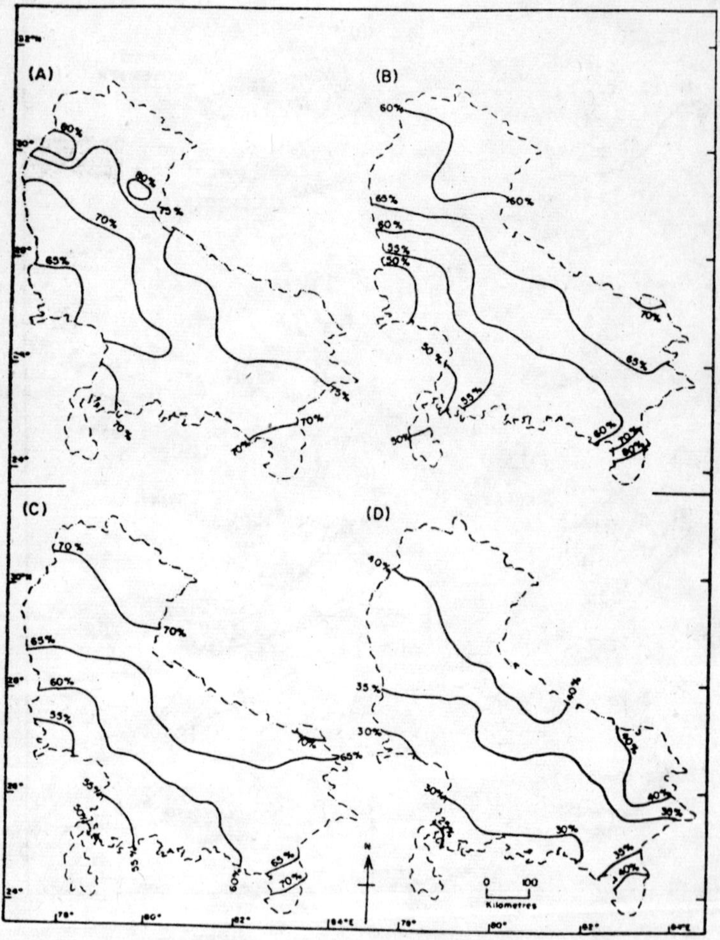

FIG. 3.3: Seasonal Distribution of Relative Humidity: (A) S.W. Monsoon, (B) Post-Monsoon, (C) Winter, (D) Hot Weather

Rainfall

Percentage of Rainfall

This is the principal rainy season during which the Ganga plain and the southern uplands of the state receive nearly 80 to 90 per cent of their annual rainfall. However, along the southern and the northern slopes of the U.P. Himalaya the season contributes 70 to 80 per cent and 50 to 70 per cent of the annual precipitation respectively (Fig. 3.4A). In the plain the seasonal percentage of rainfall is more in the east than in the west U.P. because monsoon holds over east U.P. for a longer duration.

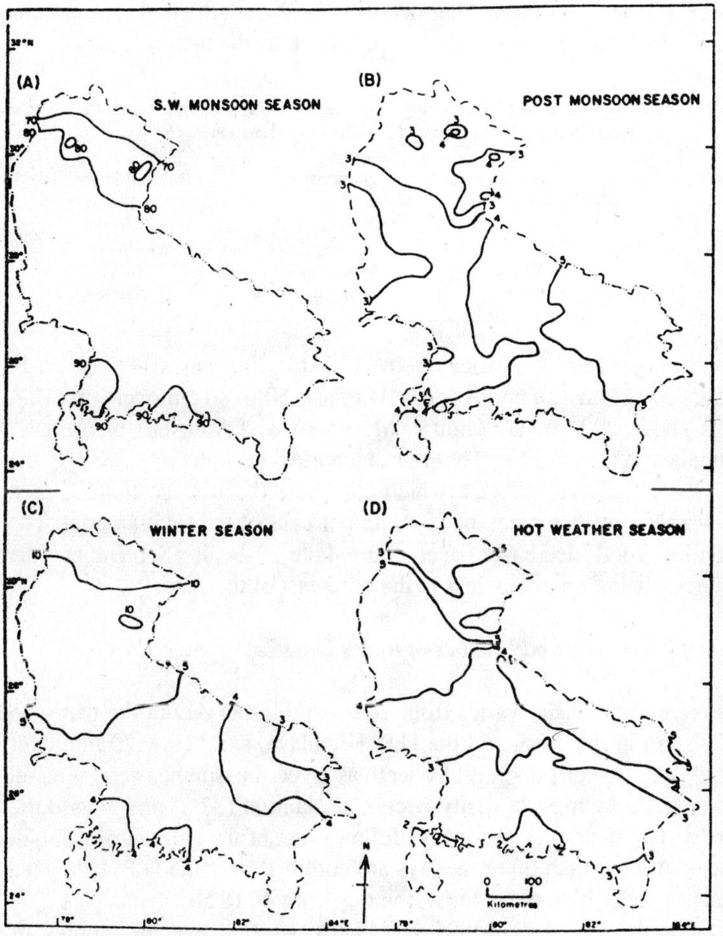

FIG. 3.4: Percentage of Mean Annual Rainfall: (A) S.W. Monsoon, (B) Post-Monsoon, (C) Winter, (D) Hot Weather

Amount of Rainfall

The special conditions under which rain occurs over the U.P. Himalaya with a marked variation in its amount is quite different from that over the plain. The southern windward slopes of the east-west trending ranges gét more rain because of the forced ascent of the monsoon wind along these slopes and the subsequent fall in temperature leading to condensation and precipitation. Therefore, we find a striking contrast in the amount of rainfall, that occurs over the southern slopes and the plains. The amount naturally decreases towards the plain.

TABLE 3.1: Rainfall in the Outer Slopes of the Himalaya Compared to that of the Neighbouring Plain during the Southwest Monsoon

Stations	Rainfall in cm	Stations	Rainfall in cm
Mussoorie	212.1	Nainital	227.3
Dehra Dun	185.7	Kathgodam	185.9
Raipur	182.8	Haldwani	175.3
Hardwar	103.9	Kitchha	110.0

Average rainfall varies from 120 to 200 cm over the U.P. Himalaya, 80 to 120 cm over east U.P. and 50 to 80 cm over west U.P. It is above 200 cm at Nainital, Mussoorie and Munsiari in the U.P. Himalaya (Fig. 3.5A). The rain increases considerably during the months of July and August which constitute the wettest months, each of which contributes nearly 30 to 35 per cent of the total amount. The monthly total decreases over the whole state in September, the decrease being relatively less in the northeast of the state.

Rainfall Intensity and Number of Rainy Days

The rainfall intensity varies from 20 to 25 mm per day in the plain, 25 to 30 mm in the Tarai and the U.P. Himalaya, and below 20 mm over Agra and its surroundings as well as over the southeastern upland. Comparatively high intensity occurs at Nainital (32.3) and Mussoorie (30.8). The pattern of rainy days follows that of the amount of rainfall. Areas of more than 60 rainy days are confined over the U.P. Himalaya except at very high elevations. It varies from 40 to 50 days in east U.P. and less than 30 days in the west. Thus, we find that the number of rainy days increases from west to east and towards the Himalayas (Fig. 3.6A).

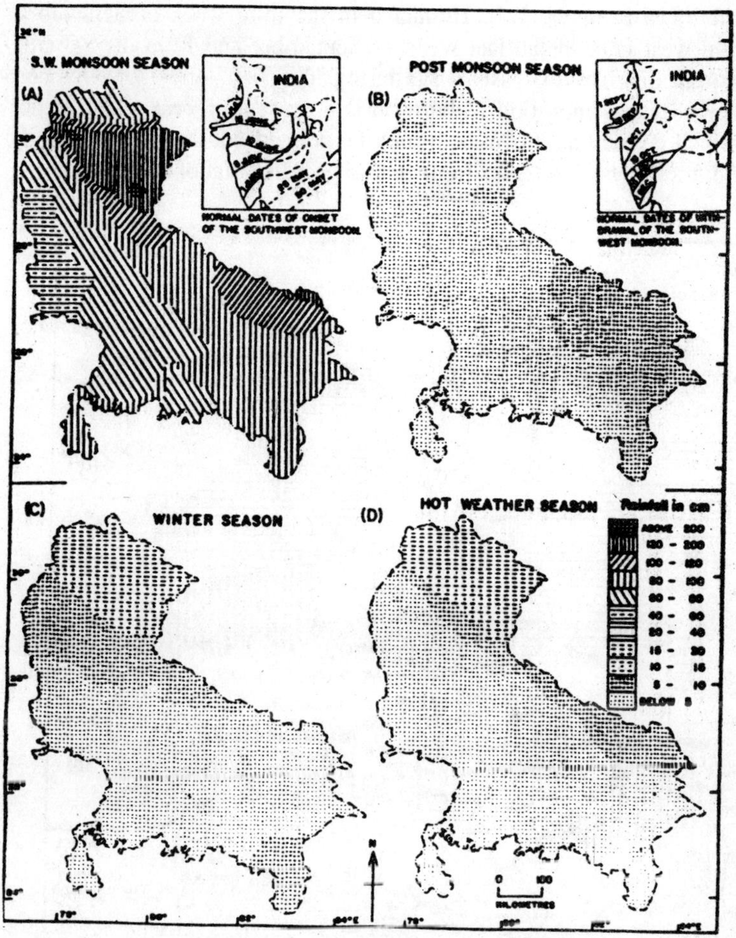

FIG. 3.5: Seasonal Distribution of Rainfall: (A) S.W. Monsoon, (B) Post-Monsoon, (C) Winter, (D) Hot Weather

Onset and Withdrawal of the Sourthwest Monsoon

It normally advances in the extreme eastern districts of east U.P. during the third week of June and into the west U.P. about a week later, but sometimes, it advances into the U.P. Himalaya a little (say about a week) earlier than into the western plains (Table 3.2). 'The standard deviation of onset of southwest monsoon is about 6 days.'[2] It withdraws from the U.P. Himalaya in the third week of September, from west U.P. in the last week of September and from the extreme eastern districts of east U.P. during the first week of October. Generally, the monsoon withdrawal is a gradual process which is not so well defined as its onset. It may be pointed out that the date of its withdrawal also fluctuates from year to year like that of its onset.

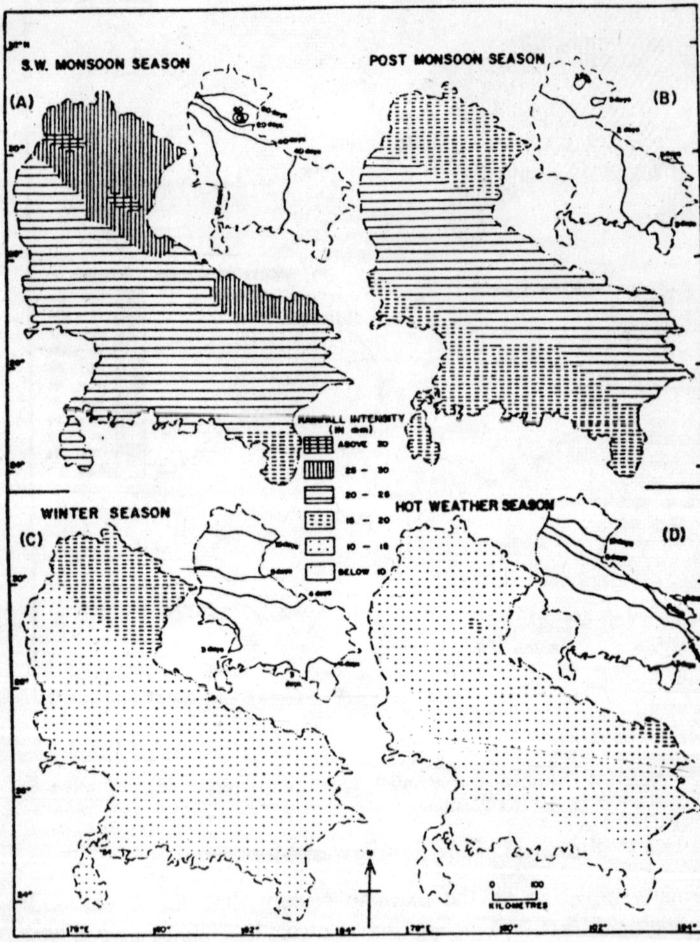

FIG. 3.6: Rainfall Intensity and Number of Rainy Days: (A) S.W. Monsoon, (B) Post-Monsoon, (C) Winter, (D) Hot Weather

TABLE 3.2: Advance of Southwest Monsoon (Dates)[3]

Area	Normal date	Earliest date	Most delayed
East U.P.	15th June	5th June	3rd July
West U.P.	25th June	10th June	9th July

Causes of Rainfall

The common synoptic situations that produce good rainfall in the state during the southwest monsoon are:[4]

- A well-settled seasonal low pressure trough along the Ganga Valley with or without a depression embedded in it.
- Appearance of depression or cyclone in the trough.
- North-south oriented trough in the western current of the southwest monsoon running northward across the western Himalaya or Tibet in the upper troposphere.

Seasonal Low Pressure Trough

The seasonal trough is the major cause of rainfall during the southwest monsoon. It extends from Punjab to West Bengal. The oscillation of this trough and its intensity determine the character of the monsoon rain with respect to its distribution and amount. It is modified by formation of depressions at the southeast end of the trough over the Bay of Bengal and over the northeast region of the Arabian Sea. During the second half of June the seasonal trough is well marked and the pressure difference south of the axis on the surface reaches a maximum of 13.6 mb (as measured by the difference in the surface pressure between Trivandrum and Delhi). By September the monsoon trough becomes weak and consequently pressure difference to the south also decreases. Towards the close of the monsoon season (during its withdrawal period), the trough adopts a more NNW to SSE orientation extending from Central Uttar Pradesh through northeast Madhya Pradesh and Orissa to the west central Bay. The following situations are possible with respect to the shifts of the trough.

(i) Monsoon trough with an embedded secondary low is an usual feature in east U.P. After the monsoon trough shifts northward, there is generally no embedded secondary low in it and the trough has a tendency to run from west U.P. to Bihar plain, resulting in active monsoon over the western parts of the state. When the axis of the monsoon trough shifts across the west U.P. it is affected by southern currents of the middle latitude westerly.

(ii) When the trough shifts south of the normal position and a depression or a cyclone is formed at the head of the Bay, the extreme southern portions of the state may have active monsoon and large part of the state may experience weak monsoon. When the trough axis shifts fairly southward sparing the state, the whole state may have little rain (particularly if the easterlies to the north of the trough are also quite strong).

(iii) When the axis of the monsoon trough shifts to the foothills of the Himalayas, it leads to break of monsoon and decrease in rainfall over the plain. This statement is not strictly correct. There are a number of occasions when the axis of the seasonal trough is towards the foothills but still the Ganga plain gets extensive rain. Such instances occur, when a trough in the western current trending north-south moves further and extends into the Ganga plain.

(iv) Seventy per cent incidence of active to vigorous monsoon occurs when the axis of the trough shifts rapidly across the area.

Depressions or Cyclones and Secondary Lows

Almost all the monsoon depressions are formed at or near the head of the Bay of Bengal (20° N) and only a few are formed in the northeast part of the Arabian Sea. Sometimes, local depressions are formed over land. The 'strengthening' and 'break' of monsoon are usually associated with the formation and movement of these depressions. In July and August only the southern part of the state is affected by depressions but its frequency is very low. In September, more depressions reach the state than in the preceding two months. Consequently in September, active to vigorous monsoon conditions prevail over the state as a result of these depressions. Very heavy rain (rainstorm) occurs over relatively small area and sometimes the spell of heavy rains continues for a few days, specially when the depressions exhibit a tendency of stagnation.

The low pressure areas also affect the state during the southwest monsoon which are remnant of depressions formed over the north Bay of Bengal or over land. They do not last for more than two days, on about 75 per cent of occasions their duration is only for one day as a low pressure area although their frequency is more in August and September than in July. Very bad weather is also associated with depressions or lows which sometimes form in the eastern parts of state and its neighbourhood.

Western Current

When the western branch of the southwest monsoon travels straight from the Arabian sea to the U.P. Himalaya through Rajasthan and western U.P., the following situations occur:

(i) When the axis of the monsoon trough moves towards the foot of the Himalayas, the easterlies over northwest India disappear and westerlies (western current) prevail extending to the upper layers of the troposphere. This leads to a 'break' in the southwest monsoon in the month of June or July for three to four days. In August and September, the duration of break may last up to a fortnight. Sometimes a long break in the later month may terminate the monsoon. During the 'break' period fluctuation in rainfall is largely determined by the movement of the western current trough over the area.

(ii) In the upper troposphere the axis of the subtropical ridge passes rapidly through 30° N (Over the U.P. Himalaya) during the monsoon. Immediately to the south of this ridge the easterlies are light. However, during the spell of strong monsoon and in the absence of any strong western current system over north India, the easterlies spread northwards and get strengthened up to 30° N. Hence, this area is liable to be affected by the western currents as well as by the easterlies resulting in spurts of rainfall. Some of the troughs in western current influence the weather even up to the 30° N latitude when the subtropical ridge (in the upper troposphere) is relatively weak.

POST-MONSOON SEASON (October, November)

This season is characterized by retreat of the southwest monsoon, clear sky with bright sunshine and rising temperature till the last week of October before falling again to reach its winter minimum. Thus, it exhibits the best weather of the year.

Temperature

Figure 3.1B shows mean monthly temperature and Fig. 3.2B the mean maximum and mean minimum temperatures in the month of November that is the transitional month between withdrawal of the southwest monsoon and onset of the winter season. Isopleths of mean temperature (actual) of 16°C and 18°C run along the foot of the Himalayas. The isopleth of 20°C is confined to the west Tarai and northern plain of the western U.P. whereas, those of 22°C and 24°C extend to the southern uplands of the state. Isopleths of the maximum temperature of 26°, 28° and 30°C and those of the mean minimum temperature of 12°, 14° and 16°C are almost similar to the mean temperature isopleths of 16°, 18° and 20°C respectively. The variations in mean, maximum and minimum temperatures are from 10° to 17.3°C, 15.5° to 25.4°C and 5.6° to 9.2°C respectively in the U.P. Himalaya (Fig. 3.2B).

Wind

During this season, surface winds are usually light over the whole state, though a few strong winds appear in October in association with cyclonic storms. Southerly and westerly surface winds prevail in Jhansi and Agra regions and northwesterly to westerly over rest of the state. Easterly surface winds also prevail occasionally in the northeast part of the state.

Relative Humidity

Mean relative humidity varies from 45.0 to 75.5 per cent during this season. Eastern part of the state has higher relative humidity (60 to 70 per cent) whereas over the southwestern and northern parts, it ranges from 45 to 55 and 55 to 60 per cent respectively (Fig. 3.3B).

Rainfall

Percentage of Rainfall

The seasonal percentage of rainfall follows the amount of rainfall which is apparent from Fig. 3.4B. It is 5 to 10 per cent in east U.P. and below 5 per cent in rest of the state with a decreasing trend towards the west.

Amount of Rainfall

The rainfall varies from 5 to 10 cm at Nainital and east U.P. and is below 5 cm in the rest of the state (Fig. 3.5B). On rare occasions, floods occur in the month of October in the U.P. Himalaya due to depressions, which terminate over there.

Rainfall Intensity and Number of Rainy Days

The rainfall intensity is from 20 to 25 mm per day in the Tarai as well as in the northern parts of the plain. The lowest intensity occurs over the Mirzapur uplands. The mean number of rainy days is 3 to 10 in the U.P. Himalaya and east U.P. and below 2 in west U.P. (Fig. 3.6B).

WINTER SEASON

Temperature

Figure 3.1C shows the mean temperature and Fig. 3.2C the mean maximum and mean minimum temperatures in the month of January. The isopleths of 12° and 14°C showing mean temperature (actual) run along the foot of the U.P. Himalaya and in the Tarai zone whereas those of 16° and 18°C in the southern and eastern parts of the state. The mean maximum isopleths of 20°, 22° and 24°C and mean minimum isopleths of 6°, 8° and 10°C trending E.W run more or less parallel to each other, in an ascending order from north to south. Variations in the mean, maximum and minimum temperatures are from 9.5° to 14.2°C, 11.6° to 18.9°C and 1.3° to 5°C respectively over the U.P. Himalaya.

Wind

In the month of January the state is often affected by western disturbances which come from the west and move towards NE and ENE direction. Winds are locally governed by these disturbances. In general, the surface winds are light to moderate and westerly to northwesterly in direction usually with speeds ranging from 2 to 5 knots per hour. Sometimes, it blows from southwest or west over the southwest parts of the upland.

Relative Humidity

The mean relative humidity is 50 to 70 per cent in the U.P. Himalaya and decreases towards the southwest coming down to less than 55 per cent in the Jhansi and Agra districts (Fig. 3.3C).

Rainfall

Percentage of Rainfall

The seasonal percentage is 10 in the U.P. Himalayas, 5 to 10 over west Tarai and northern plain of west U.P. and below 5 over west of the state (Fig. 3.4C).

Amount of Rainfall

The amount of rainfall is 10 to 20 cm in the U.P. Himalaya 5 to 10 cm in the west Tarai and northern parts of plain and less than 5 cm over rest of the state (Fig. 3.5C).

Rainfall Intensity and Number of Rainy Days

The rainfall intensity varies form 15 to 20 mm per day in the U.P. Himalaya and 10 to 15 mm in the rest of the state. The number of rainy days is 10 to 15 in the northern and 5 to 10 in the southern part of the U.P. Himalaya including west Tarai. It is below 5 over rest of the state (Fig. 3.6C).

Cause of Rainfall

During this season clouds and precipitation occur due to western disturbances the intensity of which gradually decreases from west to east. The U.P. Himalaya and west U.P. plain are usually affected by these disturbances. Occasionally they extend eastward and cause light rain over east U.P. Two or three active western disturbances on an average, visit Punjab and western Himalayas and 'break up' there in every winter month. They cause fairly widespread thundershower and snowfall in the U.P. Himalaya. Sometimes rainfall of 5 to 10 cm occurs in a day. 'Lows' from northern Arabian Sea are sometimes induced in these primary western disturbances over the south Rajasthan and move eastward up to east U.P. causing light to moderate precipitation over the plain and moderate to severe cold waves occasionally prevail. Induction of such 'lows' causes secondary disturbances to appear over north Gujarat and east Rajasthan which move eastwards and cause a few thundershower and light rain even in southern and eastern parts of the state. Sometimes a low pressure area over east central Arabian Sea is associated with an upper air closed cyclonic circulation forming over west Madhya Pradesh. When it extends to the adjoining parts of Uttar Pradesh it gives light rain.

HOT WEATHER SEASON (March, April, May)

Temperature

In March, temperature starts rising due to northward movement of the sun and high temperatures are recorded during the month of May and in the first half of June. In the month of May the mean temperature (actual) isopleths of 28°, 30°, 32° and 34°C run almost east-west the values decreasing progressively towards the north (Fig. 3.1D). Similarly, the mean maximum isopleths of 38°, 40° and 42°C and the mean minimum isopleths of 20°, 22°, 24° and 26°C also run east-west (Fig. 3.2D).

Wind

During this season winds are westerly to west-northwesterly with average speeds of 5 to 7 knots per hour in March and April, but it shows a tendency for easterly direction in May with a speed of 5 to 10 knots excepting the days of squalls or duststorms.

Relative Humidity

The mean relative humidity ranges from 20 to 50 per cent during this season. It decreases continuously from north to south as depicted in Fig. 3.3D.

Rainfall

Percentage of Rainfall

Seasonal percentage of rainfall is 5 to 10 per cent in the U.P. Himalaya and below 5 per cent in the rest of the state (Fig. 3.4D).

Amount of Rainfall

Ten to twenty cm of rainfall occurs in the U.P. Himalaya and 5 to 10 cm in the Tarai and the foothill zone. In the rest of the state rainfall is below 5 cm. It generally occurs during the month of March and spoils mango fruiting (Fig. 3.5D).

Rainfall Intensity and Number of Rainy Days

Rainfall intensity is 10 to 15 mm per day in the state excepting Agra and the southern upland (south of the Yamuna) where the intensity is below 10 mm. Number of rainy days varies from 2 to 10 days from south to north, which is depicted in Fig. 3.6D.

Other Notable Features

The western disturbances persist even in the month of March and occasionally prolong to April. Every year two or three western disturbances affect the U.P. Himalaya and sometimes the plain of the state as well. They contribute fairly widespread rain or hailstorm in March and April but rarely in the month of May. During this period the daily minimum temperature is appreciably to markedly below normal over the affected parts of the state. On the other hand, duststorms frequently occur on hot days and mitigate the air temperature, without any rainfall. They are more frequent in May and June, averaging one in a week, till the onset of the southwest monsoon. Duststorms decrease the range of visibility, as shown in Table 3.3.[5]

TABLE 3.3: Range of Visibility During Duststorms

Intensity of duststorm	Windforce on Beaufort scale	Visibility
Light	4-6	500 to 1000 m
Moderate	6-8	200 to 500 m
Severe	9 or more	Less than 200 m

Strong and hot westerly winds, locally known as *'Loo'* are characteristic feature of summer afternoons. They are experienced over an extensive area in the Ganga plain. A number of deaths occur due to dehydration of the body affected by heat strokes.

COLD WAVES AND HEAT WAVES

A region is considered by the I.M.D. to be in the grip of a moderate heat wave when the recorded maximum temperature exceeds the normal by 5° to 8°C and 'severe heat wave' if the maximum temperature exceeds the normal by more than 8°C. Similarly, a region is said to be in the grip of a 'moderate cold wave' when the recorded minimum temperature falls short of the normal by 6° to 8°C and under 'severe cold wave' if this difference exceeds 8°C.

An extract from Technical Circular Forecasting No. 23 dated the 20th December, 1958 is given below (Table 3.4) with slight modifications considering that temperature figures represent a continuous series rather than a discrete one. It gives the range of temperature along with the description associated with each.[6]

TABLE 3.4: Departure of Temperature during Heat and Cold Waves

Departure in °C	Description of change	Description of departure
–1 to 1	Little change	Nearly normal
1 to 3	Rise	Above normal
3 to 5	Appreciable rise	Appreciably above normal
5 to 8	Marked rise	Moderate heat wave
Above 8	Large rise	Severe heat wave
–1 to –3	Fall	Below normal
–3 to –5	Appreciable fall	Appreciably below normal
–5 to –8	Marked fall	Moderate cold wave
Above –8	Large fall	Severe cold wave

Cold waves are generally experienced from November to March but in some years they have also been experienced in April and October. Severe cold waves generally prevail from January to March. The monthwise maximum duration (number of days) of severe cold waves is shown in Table 3.5.

TABLE 3.5: Maximum Duration of Severe Cold Waves (in days)

Regions	Oct.	Nov.	Dec.	Jan.	Feb.	March	April
East U.P.	-	-	-	1	1	1	1
West U.P.	-	3	2	7	4	4	2

The state is rarely affected by such waves (1 per cent) in October. Frequency of cold waves is about 5 per cent in November and 11 per cent in December in west U.P., after which the frequency sharply rises to 27 per cent in January and February, when it starts affecting east U.P. The frequency 'falls to 20 per cent during March generally confined to the first fortnight of the month. During the month of April, the frequency dwindles down to 9 per cent of the total.

Cold waves occur due to the inflow of very cold air from the northern latitudes, i.e., from the extreme northwestern parts of the Indian subcontinent or even beyond, when this unusual cold air spreads over the state, it manifests as a 'cold wave'.

About 17 per cent of the severe heat waves occur in the month of March when the Sun enters the northern hemisphere. In April the occurrence of heat waves sharply falls to 7 per cent. In May their frequency increases to 10 per cent and their activity extends eastward. Nearly 54 per cent of their incidence is recorded in June when east U.P. experiences the maximum number of severe heat waves. Nearly 12 per cent of them occur in July, when their incidence are maximum in the west U.P.

The persistence of heat waves is generally 5 to 6 days. The severe heat waves persist more in June than in other months. During the heat waves, the weather is hot with clear sky providing excessive insolation.

NOTES

1. Malukar, S.L.: Note on Analysis of Weather India and Neighbourhood, *Memoirs of I.M.D.*, Vol. 28 IV (1950), p. 204.
2. International Geographical Union Regional Conference in Japan, Tokyo and NARA (28 Aug.-3 Sept. 1957), p. 592.
3. Srinivasan V. et al.: Southwest Monsoon-Typical Synoptic Weather Situation over Uttar Pradesh and Bihar, *Forecasting Manual*, III 3.5, I.M.D. (1972).
4. *Ibid.*
5. Joseph, P.V. et al.: Andhi the Convective Duststorm of Northwest India, Prepublished *'Scientific Report'*, 79/4 Table 1, I.M.D. (1979).
6. Bederar, V.C., et al.: 'Heat and Cold Waves in India', *Forecasting Manual*, I.M.D., Part IV, October (1974).

4

CLIMATIC CLASSIFICATION

CLASSIFICATION OF climate on a regional basis is necessary for its application to agriculture, forestry, human and animal ecology and other related fields. This is why many scientists have derived various climatic indices for determination of climatic types. These indices express the humidity or aridity conditions of any region, especially in the tropical countries, and they also reflect the soil characteristics as well as the vegetation types and their density. Hence, at present all attempts are to set climatic limits on the basis of vegetation regions and soil types. "The climate of any locality, however, is not created by a single climatic element but the distinctive regional combination and interrelation of several elements".[1] On account of uniformity in relief, there is a remarkable uniformity of climate over a larger area in the plains. However, variations of climate in the state are expected due to: (i) altitudinal variation in the U.P. Himalaya, (ii) varying distances from sea, (iii) encroachment of desert in the west and (iv) scattered occurrence of the northern fringe of the Deccan plateau bordering the Ganga-Yamuna plains in the south.

Mention can be made of papers by Daji (1955)[2] on the climate of Bombay state, Subrahmanyam, V.P. and Subramaniam, A.R. (1962)[3] on the climate of the Andhra Pradesh and Subramaniam, A.R. (1964)[4] on the climate of Mysore state. An attempt has also been made to delimit the climatic regions of U.P. by R.N. Tikka[5] on the basis of: (i) Thornthwaite's moisture region, and (ii) his concept of climatic division based on year to year variation of moisture index. Others who have classified climatic regions of India are Chatterjee, S.B. (1953)[6], who by improving upon Koppen's method has divided India into climatic regions, in which U.P. falls into two subregions of humid mesothermal Cwg2 and Cwg3. Spate has also mentioned only two climatic regions in U.P. (i) E-T, northern U.P., and (ii) Cwg, southern U.P., Subrahmanyam brought out a climatic divisions of India in 1956[7] based on Thornthwaite's scheme of 1948. He found four climatic divisions in U.P., i.e., Semiarid, Subhumid, Humid and Perhumid. He again classified India into climatic divisions in 1965[8] on the basis of the final scheme of Koppen and Thornthwaite. No doubt, this has exposed the presence of a number of smaller climatic regions in the state. In the present study, an attempt has been made to determine climatic indices on the basis of which U.P. can be divided into climatic regions. These indices are, Lang's Rain-Factor (1920), De Mortonne's Index of Aridity (1926), Koppen's Scheme (1936), Kira's Index of moist and dry climate (1945) and Thornthwaite's modified Scheme of Rational classification of climate (1955).

LANG'S RAIN-FACTOR (1920)

Lang, in his investigation of soil distribution, has used the rain factor, which is calculated by dividing the annual precipitation (in mm) by the mean temperature (in °C). This quotient is commonly referred to as the P-T ratio. On the basis of this P-T ratio, he classified the following three humidity provinces.

Humidity Provinces[9]	P-T Ratio
Arid	0-40
Humid	40-160
Wet	160 and above

Lang points out that 'types of natural vegetation are classified as desert, semidesert, steppe, savanna, bush and deciduous forest, and rain forest corresponding to 0-20, 20-40, 40-60, 60-100, 100-160 and 160 and above of the Rain-Factor (P-T ratio) respectively.[10] Figure 4.1A reveals that a large portion of the Ganga plain comes under steppe or savanna in the northern part and under semidesert in the southern and western parts of the plain. Rain-Factor varies in the Tarai and the U.P. Himalaya from 60 to 160 and above (except the snow clad areas).

DE MORTONNE'S INDEX OF ARIDITY (1926)

He used a slight modification of Lang's Rain-Factor, in which he suggested to divide the annual precipitation in mm by the mean annual temperature in °C + 10.[11] According to his 'Index of Aridity' the climate becomes arid, when the value of Ia is less than 20. Figure 4.1B depicts a small arid part in the state at Agra and its vicinity which is affected by the encroaching desert. In the rest of the state the Index of Aridity exceeds 20.

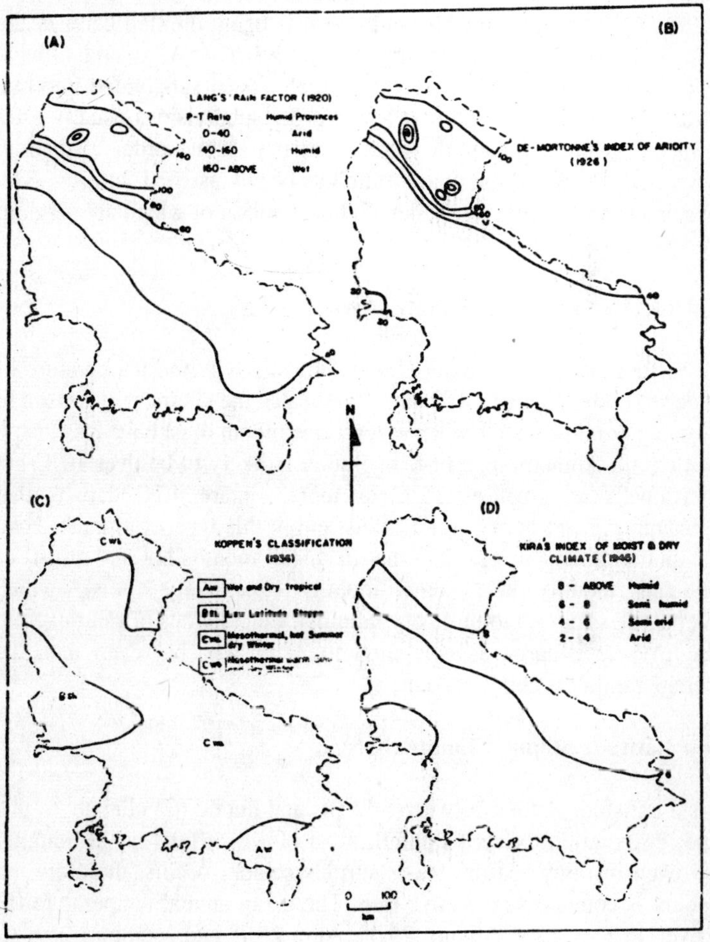

FIG. 4.1: Climatic Indices: (A) Lang's Rain Factor, (B) De Mortonne's Index of Aridity, (C) Koppen's Classification, (D) Kira's Index of Moist and Dry Climate

KOPPEN'S CLIMATIC CLASSIFICATION (1936)

A popular climatic classification of the earth by Koppen appeared in 1900 based mainly on the vegetation zones. Since then many modifications were made in his classification and the final form was presented by him in 1936. His detailed scheme is easily available in literature[12,13,14] on climatology and hence is being omitted here. In the state of Uttar Pradesh, according to this scheme, A, B and C have emerged as the major types. They have also been subdivided into two lower order divisions. The first lower order subdivisions usually refer to special feature of rainfall regime and the second order to thermal characteristics. Thus, the whole state has been classified into *Aw, Bsh, Cwa* and *Cwb* climatic types, detailed description of which are given in Fig. 4.1C.

Wet and Dry Tropical Climate (Aw)

It is also called the Savanna climate. In this type the temperature of coldest month is above 18°C and *W* indicates the distinct dry season in low sun period or winter when at least one month must have less than 6 cm rainfall. Annual range of temperature is likely to be over 10°C but occasionally as great as 15°C or more. Figure 4.1C depicts that Mirzapur and southern part of Jhansi enjoy this type of climate. Here the annual rainfall is less than that in *Af* (all months hot and moist) or *Am* (all months hot, some months excessively moist, which compensates for one to three dry months). Fluctuations of rainfall from year to year is large as compared to *Af* or *Am*, but even then the average rainfall is about 100 cm.

Low Latitude Steppe Climate (Bsh)

It is a transitional zone between arid (B) and humid (C) climate. In this type, evaporation exceeds rainfall, most of which falls in the summer and even in rainy season, water surplus seldom occurs, therefore, no amount is counted as effective rain. The mean annual temperature (h) is over 18°C, e.g., at Aligarh 25.3°, Agra 25.7° and Mainpuri 25.8°C. Annual range of temperature at these places is 19.5°, 20.2° and 19.6°C respectively.

Humid Mesothermal Climate (Cw)

It is also called 'subtropical climate'. In it the average temperature of the coldest month is below 18°C but above –3°C. *w* indicates dry winter signifying at least ten times as much rain in the wettest month of summer as in the driest month. In the Cw type found in this state, the mean summer temperature is higher (above 22°*C*) in the Ganga plain but lower (below 22°C) in the U.P. Himalaya. Hence *Cwa* applies to the former and *Cwb* to the latter.

KIRA'S INDEX OF MOIST AND DRY CLIMATE(K) (1945)

Although this method has not gained popularity in geographical literature, its application to this state as illustrated in Fig. 4.1D, appears to be quite significant. The isopleths of 8, 6, 4 and 2 correspond to the lower limits of the humid, semihumid, semiarid and arid zones.[15]

$$K = P/(W + 20); \qquad W < 100$$

Or $$K = 2P/(W + 140); \qquad W > 100$$

and $$W = \sum^{n} (t-5)$$

Where *P* represents precipitation in mm and *t*, the mean monthly temperature in °C.

This summation is carried out only for the months in which the monthly mean temperature exceeds 5°C. In Fig. 4.1D most part of the Ganga plain and the southern upland come under semiarid. The area lying between Kira's Indices of 2 to 4 embraces parts of Agra, Mathura and Mainpuri districts and represents arid type. The U.P. Himalaya enjoys humid climate, because all the indices (*K*) calculated for the area exceed 8 except that for Tehri. A strip running along the Tarai region represents semihumid type. However, the main drawback of this map is that it does not reveal any distinction between the southern upland and the Ganga plain.

As Koppen's classification of climate portrays a generalised picture of the state, the Rational classification of climate developed by Thornthwaite in 1948 and subsequently modified in 1955, has been used to give a micro level climatic divisions of the state.

THORNTHWAITE'S RATIONAL CLASSIFICATION OF CLIMATE (1955)

Thornthwaite first attempted a classification of climate of North America[16] and later that of the world. Thornthwaite's 1948 climatic classification,[17] was recognised on the basis of climatic water balance, in which he assumed moisture holding capacity of soil to be 10 cm on an average for the whole world. But Thornthwaite and Mather revised this scheme later in 1955, in which they assumed the average moisture holding capacity to be 30 cm. However, they have shown that it varies from 2.5 cm to 40 cm depending upon the soil types, its structure and type of vegetation. The relevant tables for computation of climatic water balance are given in their book published in 1957.[18]

For classification of climate of any region, on the basis of Thornthwaite's Rational classification, four relevant components, i.e., Moisture index, Thermal efficiency, Seasonal variation of effective moisture and Summer concentration of thermal efficiency must be determined. These climatic attributes are assessed with the help of climatic water balance components, i.e., potential evapotranspiration (water need), water surplus and water deficit.

The Moisture Index

Thornthwaite's expression of moisture index is $I_m = I_h - I_a$. The moisture index is used to delimit nine separate moisture regions (Table 4.1). The moisture regions and their limiting moisture indices according to Thornthwaite's 1955 climatic classification are given in Table 4.1.

Negative values of the moisture index are found in dry climates and positive values in moist ones. Thus, the moisture index value of zero, separates dry and moist climates.

Seasonal Variation of Effective Moisture

It is an important parameter to distinguish wet condition in one season and dry in another from continuously wet or continuously dry. The moisture index reflects annual quantity of moisture. It may also be used to distinguish seasonal variation. Thornthwaite has considered the seasonality of moisture condition in relation to the extent to which it reflects deficiency or surplus in each season. Koppen used *s* for summer dry and *w* for winter dry. Whereas Thornthwaite applied the letter *s* and *w* to denote seasonal deficiency in moist climates and seasonal surplus in dry climates. The details are given in Table 4.2.

TABLE 4.1: Moisture Index

Moisture regions	Climatic types	Moisture index[19]
A	Perhumid	Above 100
B_4	Humid	80 to 100
B_3	Humid	60 to 80
B_2	Humid	40 to 60
B_1	Humid	20 to 40
C_2	Moist subhumid	0 to 20
C_1	Dry subhumid	–33.3 to 0
D	Semi-Arid	–66.7 to –33.3
E	Arid	Below –66.7

TABLE 4.2: Seasonal Variation of Effective Moisture[20]

Moist Climate (A, B, C_2)		Aridity Index
r	Little or no water deficiency	0-10
s	Moderate summer water deficiency	10-20
w	Moderate winter water deficiency	10-20
s_2	Large summer water deficiency	20+
w_2	Large winter water deficiency	20+
Dry Climates (C_1, D, E)		Humidity Index
d	Little or no water surplus	0-16.7
s	Moderate winter water surplus	16.7-33.3
w	Moderate summer water surplus	16.7-33.3
s_2	Large winter water surplus	33.3+
w_2	Large summer water surplus	33.3+

Thermal Efficiency

Thornthwaite used potential evapotranspiration as an index of thermal efficiency. It is found out by temperature and day length and it expresses the amount of energy available in a climate at a particular place in terms of water that could be evaporated by this energy if water be easily available. Thus, in fact, potential evapotranspiration combines the thermal and moisture aspects of the climate. Given in the same units as precipitation, it relates thermal efficiency to precipitation effectiveness (Table 4.3).

TABLE 4.3: Thermal Efficiency

T E Index (cm)	Climatic type	T E Index	Climatic type
Climatic types according to Thornthwaite's Thermal efficiency		Seven types of Megathermal climate according to Subrahmanyam[21]	
<	E' Frost		
14.2		114.0	
	D' Tundra		A'_1
28.5		128.2	
	C'_1 Microthermal		A'_2
42.7		142.5	
	C'_2 Microthermal		A'_3
57.0		156.7	
	B'_1 Mesothermal		A'_4
71.2		171.0	
	B'_2 Mesothermal		A'_5
85.5		185.2	
	B'_3 Mesothermal		A'_6
99.7		199.5	
	B'_4 Mesothermal		A'_7
114.0			
>	A' Megathermal		

Summer Concentration of Thermal Efficiency

Seasonal variation in thermal efficiency owes to variations in temperature and day duration. In the equatorial climate where the temperature does not vary much about one quarter of energy will be received during any three months period. In the Arctic climate, in which the temperature is above freezing point for only three months of the year, 100 per cent of its thermal efficiency is received during the three summer months whereas only 25 per cent is received during the corresponding period in equatorial climate. On the basis of summer concentration of thermal efficiency, Thornthwaite has defined eight climatic types and V.P. Subrahmanyam, has further subdivided Thornthwaite's a′ type into 8 subtypes (Table 4.4).

On the basis of moisture index, seasonal variation of effective moisture, thermal efficiency and summer concentration of thermal efficiency the total possible climatic types would be 3240. In addition, V.P. Subrahmanyam has subdivided the Megathermal climate and hence this figure would increase enormously.

TABLE 4.4: Summer Concentration of Thermal Efficiency

Summer concentration (%)	Climatic type[22]	Summer concentration (%)	Climatic type[23]
	d′		
88.0		48.0	
	c'_1		a'_1
76.3		44.6	
	c'_2		a'_2
68.0		41.6	
	b'_1		a'_3
61.6		38.8	
	b'_2		a'_4
56.3		36.3	
	b'_3		a'_5
51.9		34.0	
	b'_4		a'_6
48.0		31.9	
	a		a'_7
		29.9	
			a'_8

CLIMATIC REGIONS OF UTTAR PRADESH BASED ON THORNTHWAITE'S RATIONAL CLASSIFICATION

Climatic water balance of 39 stations have been computed and with their help the state has been divided into 17 different climatic types (Fig. 4.2). Fifteen stations fall under C_1 A'_3 d a'_3 and six under D A'_3 d a'_3 climatic type. The remaining fifteen types of climate include one or two stations each. However, due to altitudinal variation, there are seven climatic types found in the U.P. Himalaya. The details of all the climatic types recognised in the state are given below:

1. A B'_1 *r* a'_2 (Perhumid, Mesothermal first with little or no water deficiency):– This type is found in the northern part of the U.P. Himalaya. The summer concentration of T E ranges between 41.6 and 44.6 per cent.
2. A B'_2 *r* a'_2 (Perhumid, Mesothermal second with little or no water deficiency):– This type is found in the northern part of Dehra Dun and Garhwal districts near Mussoorie and Mukhim. The summer concentration of T E ranges between 41.6 to 44.6 per cent.

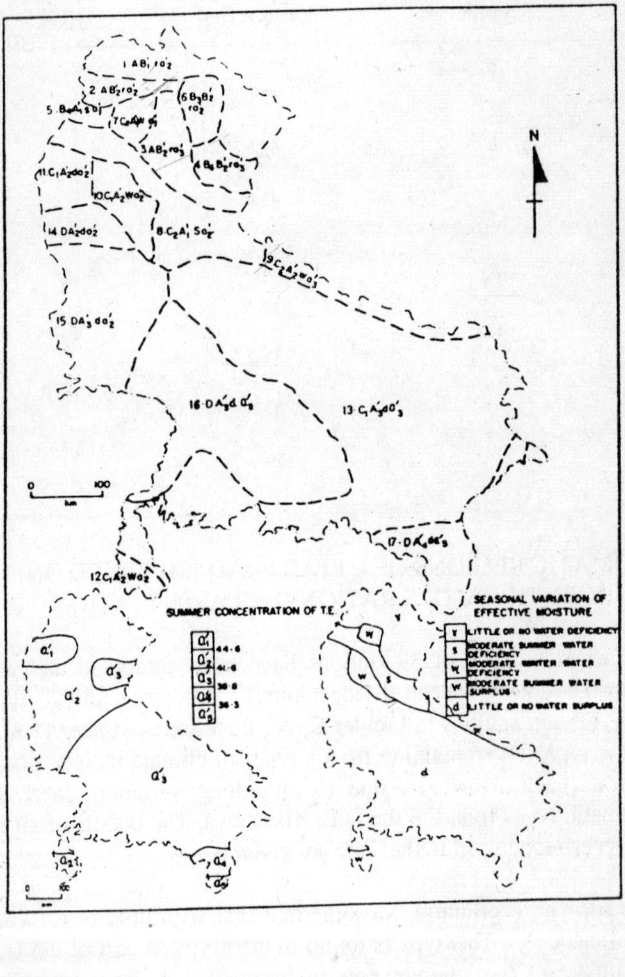

FIG. 4.2: Thornthwaite's Rational Classification of Climate

3. A B'_2 *r* a'_3 (Perhumid, Mesothermal second with little or no water deficiency):– This type is found over the northern part of

Nainital, Pauri Garhwal and parts of Almora and Chamoli districts. The summer concentration of T E ranges between 38.8 and 41.6 per cent.

4. B_4 B'_2 *r* a'_3 (Humid fourth, Mesothermal second with little or no water deficiency):– This type is found in Almora and Nainital districts. The summer concentration of T E ranges between 38.8 and 4.6 per cent.
5. B_4 A'_1 *s* a'_1 (Humid fourth, Megathermal first with moderate summer water deficiency):– This type of climate is confined to Dehra Dun and its surrounding parts. The summer concentration of T E ranges between 44.6 and 48.0 per cent.
6. B_3 B'_2 *r* a'_2 (Humid third, Mesothermal second with little or no water deficiency):– This type of climate is confined to the Joshimath area in the Chamoli district. The summer concentration of T E ranges between 41.6 and 44.6 per cent.
7. C_2 A'_1 *w* a'_1 (Moist Subhumid, Megathermal first with moderate winter water deficiency):– This type of climate is found in Tehri Garhwal district of U.P. Himalaya. The summer concentration of T E ranges between 44.6 and 48.0 per cent.
8. C_2 A'_1 *s* a'_2 (Moist Subhumid, Megathermal first with moderate summer water deficiency):– This type is found in the western part of the Tarai. The summer concentration of T E ranges between 41.6 and 44.6 per cent.
9. C_2 A'_2 *w* a'_3 (Moist subhumid, Megathermal second with moderate winter water deficiency):– This type of climate is found in the eastern part of the Tarai. The summer concentration of T E ranges between 38.8 and 41.6 per cent.
10. C_1 A'_2 *w* a'_2 (Dry subhumid, Megathermal second with moderate summer water surplus):– This type of climate is confined to Bijnor district only. The summer concentration of T E ranges between 41.6 and 44.6 per cent.
11. C_1 A'_2 *d* a'_2 (Dry subhumid, Megathermal second with little or no water surplus):– This type of climate is found in the Saharanpur district. The summer concentration of T E ranges between 41.6 and 44.6 per cent.
12. C_1 A'_2 *w* a'_2 (Semiarid, Megathermal second with moderate summer water surplus):– This type appears over the southern part of the Jhansi district. The summer concentration of T E ranges between 41.6 and 44.6 per cent.
13. C_1 A'_3 *d* a'_3 (Dry subhumid, Megathermal third with little or no

water surplus):– This type occurs in east U.P. and Bareilly district. It includes fifteen stations. The summer concentration of T E ranges between 38.8 and 41.6 per cent.

14. D A'_2 *d* a'_2 (Semiarid, Megathermal second with little or no water surplus):– It is confined to the Meerut district. The summer concentration of T E ranges between 41.6 and 44.6 per cent.
15. D A'_3 *d* a'_2 (Semiarid, Megathermal third with little or no water surplus):– This type is observed in the Agra and Aligarh districts, the driest part of the state. The summer concentration of T E ranges between 41.6 and 44.6 per cent.
16. D A'_3 *d* a'_3 (Semiarid, Megathermal third with little or no water surplus):– This type is found in Mainpuri, Etta, Ettawah, Fatehpur, Unnao, Jalaun and Hamirpur districts which include six meteorological stations. The summer concentration of T E ranges between 38.8 and 41.6 per cent.
17. D A'_4 *d* a'_5 (Semiarid, Megathermal fourth with little or no water surplus):– This type occurs in the southern upland of the Mirzapur district. The summer concentration of T E ranges between 34.0 and 36.3 per cent.

Thornthwaite's approach of climatic classification is based on moisture as well as on thermal factors with soil playing an important role in the retention and release of moisture. This approach deserves criticism mainly for the use of PE value estimated with the help of empirical law considering only mean temperature and length of day whereas some important factors like radiation, wind, humidity are not taken into consideration. PE estimates made by Thornthwaite's empirical law are thus not considered as perfect representative.

Thus, the present study deals with the determination of climatic types and their delimitation within the state as prescribed by Lang, De Mortonne, Koppen, Kira and Thornthwaite. These formulae are primarily based on two important climatic elements, i.e., rainfall and temperature on the basis of which the degree of humidity and aridity have been found out at different stations. It is noteworthy that the aforesaid scientists have classified the climatic types for the whole world. But when their schemes are applied to smaller areas like the state of Uttar Pradesh, they do not produce much significant result. Physiographically the state consists of three major units — hills, plateau and plain. The various climatic types determined on the basis of different formulae do not exactly coincide when they are superimposed over each other. But even then, the preference of Thornthwaite's approach over that of the other climatologists may be summarised by pointing out the fact that he took into consideration two other important elements, viz., potential evapotranspiration and soil moisture which have been ignored by others. These factors play a significant role in the energy exchange between the earth surface and the atmosphere which is of considerable importance in determining the type of climate.

NOTES

1. Trewartha, G.T.: *An Introduction to Climate*, New York (1968), p. 237.
2. Daji, J.A.: The Climate of Bombay State, *Journal of Indian Society of Soil Science*, Vol. 3, I.C.A.R., New Delhi (1985), pp. 133-153.
3. Subrahmanyam, V.P. et al.: Climatic Features of Andhra Pradesh, *Ibid*, Vol. 10 (1962), pp. 129-53.
4. Subramaniam, A.R.: 'Climatic and Natural Vegetation of Mysore State', *Ibid*, Vol. 12 (1964), pp. 101-12.
5. Tikka, R.N.: The Delimitation of the Climatic Regions of U.P., *Geographical Review of India*, Vol. 25, Calcutta (1963), pp. 73-83.
6. Chatterjee, S.B.: *Indian Climatology*, Calcutta (1953), pp. 263-93.
7. Subrahmanyam, V.P.: Climatic Types of India According to Rational Classification of Thornthwaite, *Indian Journal of Meteorology and Geophysics*, Vol. 7, No. 3 (1956), pp. 253-59.
8. Subrahmanyam, V.P. et al.: Koppen and Thornthwaite System of Climatic Classification as Applied to India, *Annals of Arid Zone*, Vol. 4, No. 1, Jodhpur (1963), pp. 45-55.
9. Bharucha, F.R. and Shanbhag, G.Y.: Precipitation Effectiveness in Relation to the Vegetation of India, Pakistan and Burma, *Botanical Memoirs* No. 3, Bombay University (1957).
10. Masatoshi and Yoshina, M.: Representation of Moist and Dry Climatic of Monsoon Asia According to Climatic Indices, Water Balance of Monsoon Asia, Tokyo (1971).
11. *Ibid.*
12. Bharucha, F.R. and Shanbhag: G.Y.: op cit.
13. Trewartha, G.T.: op cit., Appendix, pp. 393-99.
14. Koppe, C.F. and De Long, G.G., Koppen: *Classification of Climate, Weather and Climate*, New York (1958), pp.190-262.
15. Masatoshi and Yoshina, M.: op cit.
16. Thornthwaite, C.W.: Problem in the Classification of Climates, *Geographical Review*, Vol. 33, New York (1943), pp. 233-255.
17. Thornthwaite, C.W.: 'An Approach Toward a Rational Classification of Climate', *Ibid*, Vol. 38 (1948), pp. 55-94.
18. Thornthwaite, C.W. and Mather, J.R.: Instruction and Tables for Computing Potential Evapotranspiration and Water Balance, *Climatology*, Vol. 10, No. 3,

New Jersey (1957).

19. Mather, J.R.: *Climatology: Fundamentals and Applications*, New York (1975), pp. 112-122.
20. Mather, J.R.: op cit.
21. Subrahmanyam, V.P.: op. cit. (1956).
22. Mather, J.R.: op cit. (1975).
23. Subrahmanyam, V.P.: op cit (1956).

5

SOME MICROCLIMATOLOGICAL FEATURES

MICROCLIMATOLOGY IS the study of climate on a smaller scale. It has developed as a separate branch of meteorological science. The progressive development of a dense network of meteorological stations attracted attention towards local variations and peculiarities of climate. In mountainous regions, data gathered for stations situated along the crests, the slopes and valleys reveal their typical characteristics. Even on plain, differences were detected between stations in larger cities and those in the open country near forests or in agricultural regions.

The term 'Microclimate' has been defined as "The climatic conditions in the air layers near the ground both in the open and inside various environments".[1]

An attempt has been made here to study microclimatological differences in Uttar Pradesh especially of urban areas in the plain and on the uneven topography of the U.P. Himalaya.

URBAN MODIFICATION

The cities, no doubt, cover limited areas but are important from human point of view, because they contain concentration of large population. Urbanisation introduces radical change in the atmospheric properties of a region. Industrialization in the urban areas pollutes the air and consequently affects the resultant insolation and radiation. It also contributes extra nuclei for development of water droplets around them. The dense urban construction makes the system a better heat store, the surface more waterproof and the air to stagnate longer. A comparison of climatic data of an urban centre through the urban fringe with those of open rural environment will clearly reveal the progressive contrast between the two extremes. Such urban-rural comparison can at best reveals only an approximation of urban modification.

Temperature

If the air temperature of an urban region be compared with that of the surrounding area on a micro level, we get a glimpse of an urban heat island as depicted in Fig. 5.1A.[2] The portion of graph coinciding with urbanized area looks like an island in a vast sea of rural environment. This island reveals the relief feature of a plateau with its peaks coinciding with the heart of the city and plunging around the urban periphery depicting a cliff-like feature coinciding with the urban fringe. However, if a city is surrounded by industrial estates detached from the main city, the picture will be different from what has been described above. In this case, the heat island will appear to be surrounded by a chain of small detached islands. No special study has been made of any particular city of the state, but personal observations in Varanasi, for a period of about fifteen years, justify this concept of urban heat island. The peak of heat island is observed near the Chowk and its surroundings. Whereas, the cliff is encountered near the Assi bridge on the south, beyond Manduadih on the west, along the outer fringe of Shivapur, Pandeypur and Ashapur crossing on the north and along the bank of Ganga on the east. The ameliorating effect of the perennial Ganga is observed along its bank throughout the year in Varanasi. It is usually not expected in cities which are deprived of such river bank site.

The densely constructed *'Pucca* buildings and metalled roads, capable of capturing more insolation, are responsible for converting the urban centre into a heat store. Even after sunset, when excessive radiation occurs much of the heat is arrested within the complicated structures and is allowed to escape only partially. The multistoreyed buildings and the streets layouts affect the wind direction as well as its speed. It is needless to emphasize that the forenoon and afternoon sun affects discriminately the eastern and the western faces of the building. The northern and southern faces are also not affected equally because of the oscillation of the sun. Thus, we find that the urban heat island morphology is strongly controlled by the varying character of individual city.

City contains a large quantity of fine dust, CO_2, SO_2 etc. in surrounding industrial areas. Their effects are remarkable especially when the wind speed is low and the inversion layer is fully developed.

The smoke absorbs long wave radiation and exerts a blanketing effect on insolation during day time and the upward movement of radiation during night. Consequently, the maximum temperature becomes a little lower and the minimum temperature fairly higher in city.

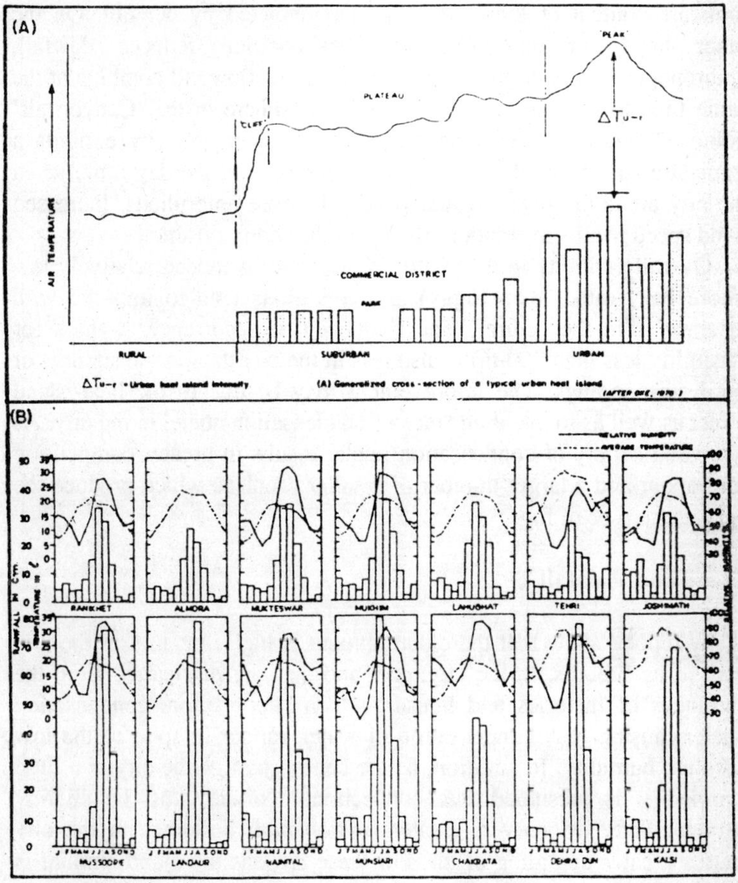

FIG. 5.1: (A) Urban Heat Island; (B) Variation in Temperature, Relative Humidity and Rainfall in the U.P. Himalaya

Humidity and Fog

"Urban-rural humidity differences are rather small and the spatial pattern is often complex. During the day, rural humidities are higher and this may be attributed to the greater rural evapotranspiration. In the early evening, the rural air cools more rapidly and becomes more stable than the urban canopy air. Therefore, the rural humidity decreases through the night, a vapour inversion forms, and the moisture content of the lower layers is depleted by dewfall. On the other hand, in the city, weak evaporation reduces dewfall, anthropogenic vapour and the stagnation of air flow, all combine at the same time to maintain a more humid atmosphere in the 'Canyon air' volume."[3] During night time humidity excess in the city exhibits a 'moisture island' similar to that of temperature. In the day time, when the city air is drier, the spatial pattern is more amorphous. Increased wind speed results in smaller urban-rural humidity differences.

Overall visibility in the centre of large city is indeed relatively low (more hazy rather than foggy) and conditions tend to improve with distance from the centre. "The frequency of occurrence of thick for (visibility less than 200 m) is also less in the city than in the suburbs or rural surroundings. The improvement may be due to the heat-island effect as well as to the abundance of condensation nuclei in the city. As increased supply of condensation nuclei results in greater competition for vapour and a larger number of smaller droplets, which produce the very dense type of fog."[4]

Cloud and Rainfall

Many reports show that the cloud amount is high in centre of the city than in its suburbs. In the city, soot and dust are discharged from the chimneys of factories and houses, which supply more condensation nuclei causing easy condensation of water vapour in spite of the low absolute humidity. In addition, as the central part of the city is a little warmer, it is presumed that convectional current must be slightly stronger there. Moreover, greater friction with buildings in the city causes greater upgliding of air. For these reasons the cloud amount is greater in the city, and as a result, the precipitation amount becomes greater. Although increased precipitation does occur in urban areas, it takes time for the vapour and the nuclei to be carried up to condensation level for the droplets to form and grow to a raindrop size. On an annual basis the total receipt of the down wind enhancement of precipitation in an urban area is 5 to 30 per cent above that of the surroundings. It is also noticed that the number of drizzly days increase in accordance with the development of city.

Thus, the most important cause for special climatic characteristics in city is air pollution, which is the main source of condensation nuclei resulting in city fog and increasing drizzly days. It also plays its part in causing green-house effect and in lowering the intensity of insolation.

THE UNEVEN TOPOGRAPHY OF THE U.P. HIMALAYA

Similarly, temperature, relative humidity and rainfall (Fig. 5.1B) in the U.P. Himalaya depends on type of terrain (inter-montane valleys, slopes and crests) on one hand and their aspects (windward or leeward slopes) on the other. The large angle of incidence of rays sunshine along the southern slopes throughout the year causes higher temperature on these slopes in comparison to the northern ones. The more or less east-west trend of the ranges intercept the monsoon winds which rise along its southern slopes causing adiabatic cooling of the ascending air parcel and consequent condensation of moisture.

Temperature

Altitude has the main factor in controlling temperature (high temperature in the deep valleys but quite low in the region of permafrost) but, at all elevations in the summer the effect of sun's rays is excessive. Dry air is heated only at the bottom and it loses on an average 0.98°C (D.A.L.R.) for every 100 m of ascent. In a diabetic cooling of saturated air (S.A.L.R.) reduces the lapse rate falls because of the release of latent heat in the process of condensation. The actual lapse rate, as observed, both in the valleys as well as on mountain slopes is less than that calculated theoretically because even dry ascending air is warmed to some extent and the isothermal surfaces get disturbed bending upwards near the slopes, while saturated air has a considerable absorbing power.

Distribution of Temperature

As every latitude has its characteristic mean annual temperature, variation in altitude has also a distinctive trend of annual and daily ranges of temperature. Table 5.1 gives the mean monthly temperature of 17 places.

TABLE 5.1: Mean Monthly and Annual Temperature (°C) in the U.P. Himalaya

Stations	Elevation(m)	Jan.	Feb.	Mar.	Apr.	May	June	July	Aug.	Sept.	Oct.	Nov.	Dec.	Annual	No. of years
Mussoorie	2266	6.4	7.8	11.7	16.5	20.0	20.3	18.4	17.9	17.1	14.9	11.6	8.5	14.2	30
Dehra Dun	680	12.6	14.8	19.4	24.6	28.9	29.5	26.8	26.1	25.5	22.2	17.5	14.0	21.8	30
Mukteswar	2311	5.8	7.3	11.1	15.4	18.7	19.0	17.6	17.3	16.5	14.2	11.2	8.2	13.6	30
Nainital	2034	6.0	8.6	12.3	15.8	19.7	20.5	19.2	18.8	17.9	14.2	10.7	8.3	14.3	10
Joshimath	1861	6.8	8.4	12.3	17.0	20.0	22.2	21.1	20.8	19.5	17.0	11.3	8.8	15.4	10
Tehri	770	12.0	14.5	19.2	24.4	28.3	29.8	28.2	28.1	27.0	20.3	17.3	12.9	21.8	3
Mukhim	1800	7.8	8.8	12.7	17.4	20.5	21.4	20.4	20.3	19.4	16.4	12.6	9.9	15.6	7
Munsiari	2300	6.5	8.3	8.4	13.8	15.4	18.9	18.2	17.4	16.5	12.7	10.6	8.5	13.1	3
Landaur	2289	3.2	6.7	9.5	13.5	17.3	20.3	18.5	17.7	17.7	13.6	9.7	6.7	12.9	5
Chakrata	2149	5.3	6.5	10.2	15.3	18.1	20.0	18.1	17.8	17.2	14.2	11.2	8.1	13.5	11
Ranikhet	1850	7.8	9.6	13.5	18.4	19.9	21.8	19.9	19.6	19.1	16.1	13.2	10.1	15.8	9
Almora	1676	7.9	11.2	14.1	18.2	21.3	23.9	22.9	22.5	22.5	18.6	14.4	10.7	17.3	7
Lahughat	1722	6.9	7.7	11.3	16.1	18.9	21.7	21.7	21.5	20.4	17.3	11.1	8.0	15.2	3
Pauri	1631	—	—	—	15.6	20.6	22.8	22.2	20.6	19.4	16.1	—	—	—	1
Srinagar	594	—	—	—	21.6	26.7	28.9	26.1	28.9	26.1	—	—	—	—	1
Kalsi	609	14.6	16.1	17.0	25.4	27.3	30.0	28.7	26.8	25.1	21.6	17.3	15.4	22.1	1
Niti	3494	—	—	—	—	—	—	14.4	14.4	12.8	—	—	—	—	1

The mean monthly temperature varies from 3.2°C at Landaur (2289 m) to 14.6°C at Kalsi (609 m) in the coldest month of January and from 18.9°C at Munsiari (2300 m) to 30.0°C at Kalsi in the hottest month of May, showing the effect of altitude. When comparison is made even between almost equal altitudes laying behind the outermost slopes on oneside and along the southern outskirts on the other, we find temperature variation. For example, Mussoorie is too hot in comparison to Chakrata. In the intermontane deep valleys, for example, at Joshimath, the temperature is high during April to October (from 17.0°C to 22.2°C) due to reflection from the two steep hill slopes which enclose the valley on either sides. At other places in the valley, like Srinagar and Tehri, the monthly temperature varies from 12.0° to 29.8°C from the coldest month of January to the hottest month of June. From 150 m to 300 m above the valley floor, the climate is more equable, although at almost all places below 1500 m heat is excessive in the months of May and June and above 1500 m the shade heat in summer is sufficiently reduced due to the effect of altitude.

Another notable feature is that the amount of insolation received along the mountain slopes differs according to the gradient and direction of slopes. South and southwest facing slopes are supposed to receive maximum effective insolation than the north and northeast facing slopes, because the sun remains south of this area throughout the year.

Table 5.2 gives the mean monthly excess temperature of the solar thermometer above the maximum in shade at Chakrata, Ranikhet and Dehra Dun. The table shows clearly that the excess temperature is greater at higher elevations in comparison to lower altitudes.

Range of Temperature

Both diurnal and annual ranges of temperature decrease on ascending from the plain up to the elevations ranging from 2100 to 2400 m, beyond which they again increase, their greatest values being attained at the highest elevations. This is probably due to more rarified air at higher elevations. Table 5.3 shows that the least annual and diurnal ranges of temperatures are found at Mussoorie, Mukteswar and Nainital which are situated above 2000 m. The driest month of April and May have largest diurnal range and most humid months of July and August have the least. There is a secondary minimum of temperature range coinciding with the maximum of humidity in month

TABLE 5.2: Monthly Mean Excess Temperature (°C) of the Solar Thermometer above the Maximum in Shade

Stations	Elevation(m)	Jan.	Feb.	Mar.	Apr.	May	June	July	Aug.	Sept.	Oct.	Nov.	Dec.	Annual
Dehra Dun	680	10.6	11.2	14.2	13.9	13.8	12.3	12.0	11.2	14.3	14.7	13.1	11.0	12.7
Chakrata	2149	22.8	20.6	21.9	31.9	21.7	19.1	17.2	16.4	21.3	21.0	21.9	20.3	19.9
Ranikhet	1850	16.2	15.9	17.0	15.8	15.7	13.9	10.4	11.2	15.4	14.1	11.5	13.2	14.2
Roorkee	270	11.2	10.8	13.3	12.6	12.1	10.7	11.1	11.0	12.5	12.5	12.8	11.4	11.8
Bareilly	173	10.0	14.4	12.3	12.4	12.9	12.3	10.4	10.1	9.6	11.1	11.3	9.4	11.1

TABLE 5.3: Mean Diurnal and Annual Ranges of Temperature (°C) in the U.P. Himalaya

Stations	Mean Diurnal Range													Mean Annual Range
	Jan.	Feb.	Mar.	Apr.	May	June	July	Aug.	Sept.	Oct.	Nov.	Dec.	Annual	
Mussoorie	7.7	8.2	9.0	9.3	9.7	7.7	4.9	5.6	5.6	7.6	8.4	8.4	7.6	13.9
Dehra Dun	13.0	13.2	14.0	15.1	14.7	11.7	7.3	6.8	8.3	12.1	14.4	13.9	7.1	16.9
Mukteswar	7.8	8.3	9.4	9.4	9.8	8.4	5.8	5.6	6.8	8.5	9.2	9.7	8.1	12.3
Nainital	8.7	9.0	8.8	10.9	8.6	7.1	4.5	4.7	6.1	8.7	10.1	9.6	8.0	14.5
Joshimath	9.6	9.9	11.2	10.2	11.7	10.4	6.9	6.9	8.0	11.7	10.5	9.8	9.9	15.4
Tehri	13.9	15.6	16.1	17.2	17.1	14.8	9.7	9.2	10.6	18.4	16.2	16.0	14.6	17.8
Mukhim	8.3	9.0	10.2	11.0	10.1	8.9	6.0	6.0	7.3	9.3	9.9	9.0	8.7	13.6
Munsiari	9.3	8.9	8.8	10.0	11.0	9.2	5.9	5.6	8.4	7.8	9.8	8.3	8.6	12.4
Landaur	8.5	10.4	10.1	10.9	11.1	11.8	7.3	5.3	6.8	6.1	9.3	10.1	9.1	16.0
Chakrata	10.2	10.2	10.0	10.9	10.8	9.1	7.0	6.3	6.7	9.6	10.0	10.2	9.2	14.7
Ranikhet	9.0	9.3	10.1	10.3	10.0	8.3	6.2	6.1	7.3	8.5	9.5	9.0	8.6	14.0

of January and a secondary maximum in the dry and cloudless month of November. For example, the diurnal range of temperature rises up to 7.7° and 9.7°C, 8.3° and 9.8°C and 9.0° and 10.3°C at Mussoorie, Mukteswar and Ranikhet respectively in coldest and hottest months in a year. On the other hand, stations situated in valleys like Tehri, Joshimath and Dehra Dun the daily range of temperature goes up to 13.9° and 17.2°C, 9.6° and 11.7°C, and 13.0° and 15.1°C in the coldest month of January or February and hottest month of April or May. The annual and daily range of temperature between hottest and coldest months are also greater over the interior than on the outer ranges and valleys owing to the larger proportion of cloudy sky and greater humidity of the air in the latter region.

Relative Humidity

Like temperature, relative humidity also differs considerably between places situated almost at equal elevations along the outer and inner ranges as well as at different elevations. It also varies greatly in accordance with the wind direction. Along the southern slopes of the Himalayas, the monsoon winds are forced to ascend which leads to their adiabatic cooling and condensation of moisture. Thus, we find high percentage of relative humidity on the southern slopes. On the leeward side, the temperature rises and humidity decreases. Therefore, high humidity is observed at Mussoorie, Nainital and Chakrata, varying from 85 to 95 per cent during the monsoon months. At other places on the leeward side it varies from 70 to 85 per cent during the same period. Relative humidity is greater over the forest-clad slopes and valleys in comparison to the steep and bare mountain slopes. Ranikhet and Dehra Dun are too moist in comparison to Almora and Tehri. It is apparent from Table 5.4. The relative humidity is also high at places situated near the watershed. For example, Nainital, influenced by the lake, is more moist than Ranikhet. The relative humidity of the air at all elevations up to 3600 m presents double maxima, one maximum occurring in the mid-cold season and the other in the mid-rainy season. However, at all elevations the highest relative humidity is reached in August. In the cold weather the sub-Himalayan and valley stations are more humid than the upper hill stations because the air is cooled by radiation during the night and the cold air settles down in the plain and valleys. In this season the air also appears to be drier at the hill stations in the morning than in the evenings.

TABLE 5.4: Approximate Mean Relative Humidity at Some Stations in the U.P. Himalaya

Stations	Jan.	Feb.	Mar.	April	May	June	July	Aug.	Sept.	Oct.	Nov.	Dec.	Annual
Mussoorie	65	64	54	45	45	67	94	96	90	71	59	60	68
Dehra Dun	69	59	48	34	34	51	81	85	77	65	66	69	61
Mukteswar	56	56	46	38	43	65	91	93	87	64	50	48	61
Nainital	66	51	50	37	42	63	87	87	81	67	62	64	62
Joshimath	47	50	42	41	38	62	82	84	84	64	48	48	55
Tehri	65	67	54	37	35	43	71	78	72	61	67	69	60
Mukhim	58	52	49	41	41	59	89	87	84	69	55	53	60
Munsiari	48	49	35	27	34	59	84	91	82	73	46	43	64
Chakrata	68	64	56	49	57	68	94	94	89	64	49	58	67
Ranikhet	60	58	52	39	49	61	85	86	79	60	53	54	62

Rainfall

The amount of rainfall varies with elevation as well as with the situation of the stations concerned. The average monthly and annual rainfall of some places has been classified into four groups according to their position, i.e., (a) near the foot zone of the U.P. Himalaya, (b) on the southern slopes of the ranges, (c) on the northern slopes of the ranges and (d) in the intermontane valleys (Table 5.5). The maximum amount of rainfall appears on the middle part of the windward slopes. At elevations with temperature much below the dew point, the decrease in absolute humidity reduces the amount of rainfall. The average rainfall on the southern slopes of the ranges in the Kumaon and Garhwal are approximately as indicated in Table 5.6.

Similarly, there is a sharp contrast in the amount of rainfall along the windward and the leeward slopes of the Himalayan ranges. This orographic rain is due to the high adiabatic lapse rate and the resulting high relative humidity. The leeward slopes, on the other hand, get little rain because of the katabatic heating of the wind causing its rain shadow effect (Table 5.7).

The summer rain gradually diminishes in amount as we move along an Himalayan range from east to west as well as along the high and continuous ranges situated north of the outer range. Consequently as we approach the Indian watershed, the quantity of precipitation is very small, the monsoon only just causing a few partial snowfall. For example before reaching the greater Himalaya the monsoon has to cross the Dhaula Dhar range, which checks their advance. Only some clouds are able to cross the Dhaula Dhar and even they fail to reach beyond the Main Range. Hence, there are marked differences in the amount of rainfall in the front and rear of the Main Range. From June to September the average rainfall is 37 to 50 cm along the front slope and 20 to 25 cm on the rear slope. The winter depressions cause 3 to 5 m of snowfall from November to May and sometimes it continues up to mid-June. In the region behind the great range the summer precipitation is less than 25 cm. because of high altitude even the summer rain is accompanied by snowfall, blizzards and hails. In winter snowfall also occurs at elevations between 1500 and 2400 m south of the Greater Himalaya. During the summer season, the snowline retreats to an altitude of about 6000 m.

TABLE 5.5: Rainfall in the U.P. Himalaya (in cm)

Stations	Altitude(m)	Jan.	Feb.	Mar.	April	May	June	July	Aug.	Sept.	Oct.	Nov.	Dec.	Annual
(a) *Foot zone of the Himalaya*														
Dehra Dun	680	5.78	6.67	3.78	1.96	3.58	18.41	65.56	71.29	30.46	4.20	0.74	2.49	214.92
Kalsi	609	5.61	8.69	5.64	2.44	3.91	20.37	63.73	68.35	25.50	2.26	0.10	2.51	209.11
Katdwara	339	4.19	4.20	2.15	1.54	2.54	17.25	55.70	54.18	24.59	3.70	0.63	1.92	172.59
Ramnagar	360	4.11	4.02	2.01	1.03	3.30	16.45	44.72	45.94	20.89	3.08	0.48	1.48	147.51
Haldwani	348	4.14	4.86	2.13	1.23	3.51	24.42	62.16	61.19	27.55	3.63	0.32	2.00	197.14
(b) *Southern slopes of the Himalaya*														
Chakrata	2149	3.58	7.75	8.66	4.78	7.95	20.24	43.76	41.43	15.09	1.55	0.33	2.95	158.07
Mussoorie	2166	7.92	8.97	6.63	3.72	5.48	20.85	78.22	80.06	32.94	4.11	1.26	3.56	253.72
Landaur	2289	6.27	3.66	4.27	6.32	11.68	18.39	67.08	78.38	19.53	1.57	1.73	2.41	221.29
Nainital	2024	6.98	7.31	5.26	3.80	8.48	39.08	76.94	75.01	36.27	6.10	1.29	2.54	269.06
Munsiari	2300	5.30	7.20	8.86	4.91	5.70	27.87	79.70	71.80	36.32	4.06	1.86	1.80	255.38
Mukteswar	2311	5.69	6.19	4.85	3.64	5.63	17.57	31.62	34.66	20.46	4.35	0.90	2.46	138.02
(a) *Northern slopes of the Himalayas*														
Lahuaghat	1722	6.80	7.60	5.00	3.00	6.20	19.50	33.90	28.70	19.80	4.40	1.20	2.50	138.60
Almora	1676	4.29	4.93	4.21	2.80	4.84	14.37	26.49	23.43	13.03	3.22	0.74	2.15	104.50
Ranikhet	1850	5.29	6.21	4.63	3.11	5.01	14.11	32.49	34.44	16.59	3.36	0.73	2.29	128.26
Pithoragarh	1695	4.44	5.63	6.87	2.82	7.26	18.28	29.97	28.72	14.95	3.32	0.65	2.17	125.08
Pauri	1631	6.07	8.30	5.51	3.22	5.16	13.25	32.60	36.00	14.84	3.39	0.81	2.82	131.97
(b) *Intermontane valleys*														
Tehri	770	6.38	5.67	4.47	1.97	3.49	5.05	30.15	31.04	21.82	2.04	0.73	1.73	114.54
Srinagar	598	5.52	5.62	3.69	2.21	4.26	10.82	24.13	23.56	9.74	2.50	0.56	2.16	94.77
Mukhim	1800	7.68	5.24	5.79	3.71	7.04	14.51	50.34	35.33	25.15	4.32	0.53	5.21	164.85
Joshimath	1861	7.13	9.07	9.65	5.33	3.53	9.71	17.28	18.23	11.35	2.81	1.22	2.77	98.08
Okhimath	1518	6.47	6.63	5.16	4.08	7.34	19.64	54.42	55.00	23.24	3.58	0.60	2.20	188.36

TABLE 5.6: Average Rainfall at Various Elevations in the U.P. Himalaya

Elevation from the mean sea level (m).	Rainfall (cm)	Elevation from the mean sea level (m)	Rainfall (cm)
250	108	2100	220
300	150	2400	130
600	312	2700	68
900	398	3000	35
1200	410	3300	18
1500	372	3600	10
1800	305		

Source: Based on the Himalayan Gazetteer, Vol.1, Part 1, 1971, p. 254.

TABLE 5.7: Rainfall on the Windward and Leeward Slopes in the U.P. Himalaya during the Southwest Monsoon

Stations on the windward side	Rainfall (cm)	Stations on the Leeward side	Rainfall (cm)
Okhimath	152.3	Joshimath	134.3
Nainital	259.8	Ranikhet	122.6
Mussoorie	266.8	Almora	98.9
Rajpur	266.5	Tehri	104.3
Pauri	127.6	Srinagar	93.0

On the basis of the aforesaid discussion we may conclude that presence of considerable absolute and relative relief in a mountainous area causes a lot of variation in the spatial distribution of climatic elements. However, paucity of data, specially in the snowclad and inaccessible rugged terrain, has stood in the way as a result of which full analysis in the present study has been hampered.

NOTES

1. Venkatanarasinga Rao, M.B. et al.: A Brief Note on Microclimatic Studies at Coimbatore, *Indian Journal of Meteorology and Geophysics*, Vol.4, No.2 (1953), p.184.
2. Quoted from Oke, T.R.: *Boundary Layer Climates* (1978), p. 254 (Fig. 8.13).
3. Oke, T.R.: *Boundary Layer Climates*, New York (1978), p. 260.
4. *Ibid.*, p. 262.

6

POTENTIAL EVAPOTRANSPIRATION: Its Determination and Distribution

POTENTIAL EVAPOTRANSPIRATION is important for computing water balance. It is the upward movement of water vapour and thus represents water loss in the hydrological cycle. It comprises of two distinct processes: (i) evaporation from soil, water bodies, snow surface and intercepted precipitation, and (ii) transpiration representing the volume of water use in the vegetation growth for building of plant's tissues and finally its escape to the atmospheric air through the leaves. Hence, evapotranspiration means evaporation plus transpiration in any specified time. It is complex process, which is partly physical and partly physiological. "Physically evapotranspiration depends upon the solar energy, wind movements, vapour pressure gradient and the movement of water in the soil either in liquid or in vapour state. Solar energy, which supplies the latent heat of vaporization for converting water into vapour, varies with the time of the day, season of the year, latitude, and altitude, while the movement of water in the soil depends upon the structure, texture, temperature, state of wetness and concentration of the soil solution. Physiologically the plant is involved from the lowermost root hair to the topmost stoma. Hence, to understand this complicated phenomenon, which is of vital importance in agriculture, meteorology, hydrology, irrigation and forestry, it is necessary to know the relation between soil water on the one hand and the micro-meteorological factors on the other hand and finally the inter-relation between these two groups."[1]

The methods by which evapotranspiration can be determined under natural conditions are: (i) the turbulence method, (ii) the heat balance method, and (iii) the volumetric method. The one method so far developed for measuring actual evapotranspiration from a field or any other natural surface without disturbing the vegetation cover in any way, is called 'vapour transfer' method. Water vapour, when it enters the atmosphere from the ground or from plants is carried upward by the moving air in small eddies or parcels of air, that are replaced by drier eddies from above. It is well known that when evaporation is taking place, the amount of moisture present in the air is greatest near the ground and decreases with height above it. If the rate at which the air near the ground mixes with that above is determined and the difference in the water vapour contents at the two levels is measured, we are in a position to find out both the rate and amount of evapotranspiration. Further more, the amount of water condensed as dew can also be determined.

This method is not easy either to understand or to use. It requires measurements more precise than are usually made. Moreover, the coefficient of turbulent transfer of air is not constant, varying from time to time and from place to place and height to height. In spite of these difficulties, the method can be perfected and will answer many questions for climatology and meteorology.

THE CONCEPT OF POTENTIAL EVAPOTRANSPIRATION

The concept of potential evapotranspiration was first popularised by Thornthwaite[2] who defined it as the water loss from surface completely covered with vegetation and supplied with abundant water at all times. The quantity of P.E. depends only on the amount of solar energy received by the surface and its resulting temperature. Actual evapotranspiration is the actual water loss. It will be equal to the potential when the water supply exceeds or equals the water need. When the supplied water into the soil falls short of the water need, then A.E. can be estimated as precipitation plus the quantity lost from soil moisture. Soil moisture utilization occurs as long as the soil moisture content lies between the field capacity and the wilting point, on the other hand, there will be deficit if the moisture content is less than the wilting point. Thornthwaite gave a comprehensive statement that A.E. = P.E. as long as there is requisite available water and A.E. = O at moisture contents below the wilting point or at temperature below the freezing point. At or near field capacity, evapotranspiration will occur at approximately the potential rate, but when the moisture content is reduced by half, the evapotranspiration will also take place at half the potential rate. Penman[3] defines potential transpiration as the amount of water that transpires in any unit of time by a short green crop, completely shading the ground, of uniform height and never short of water: He writes that for complete covers of different plants, having about the same colour, i.e., the same reflection coefficient, the potential transpiration rate is the same irrespective of plant or soil type. Thus, potential evapotranspiration has been defined in several ways, all of which have essentially the same meaning.

DETERMINATION OF POTENTIAL EVAPOTRANSPIRATION

Thornthwaite's Method

Thornthwaite first devised to determine the potential evapotranspiration by means of an empirical formula involving mean monthly temperature and relative humidity. He refined this formula to achieve more satisfactory results which were utilized in many regions of the world. Two main reasons for its wide use are that the method is simple and that it requires only mean temperature data and latitude (to eliminate all other factors). He first found out an equation to obtain heat index I, derived as follows.[4]

Heat index $i = (t/5)^{1.514}$. The sum of the 12 monthly values thus derived is called heat index I. Where t is mean monthly temperatures in °C. The final form of measuring the potential evapotranspiration was obtained by the following equation:

$$U = 1.6\,(10\,t/I)^{a}$$

Where, U is unadjusted potential evapotranspiration or consumptive use for a month in cm

$$a = 0.000000675\,(I)^3 - 0.0000771\,(I)^2 + 0.01792\,I + 0.49239$$

This equation, however, gives only unadjusted rates of potential evapotranspiration. At temperatures over 26.5°C the relation between temperature and unadjusted P.E. is the same every where. But at lower temperatures there is an increasing divergence in the relationship between temperature and unadjusted potential evapotranspiration. These unadjusted rates of P.E. are required to be increased or reduced by multiplying it with a factor which varies with month and latitude.

Thornthwaite's formula completely lacked in mathematical elegance. It was very complicated and without monograms and tables as computing aids, and thus it was quite unworkable. Hence, he prepared tables[5] and evolved out a new technique for computation. The first step is to obtain the heat index *I*, which is the sum of the 12 monthly heat indices *i*. These monthly indices are obtained if the monthly temperature of a station is known. The next step is to find the unadjusted P.E. values for each month. Finally these values are adjusted if these unadjusted potential evapotranspiration are multiplied by monthly correction factors. It depends on the latitude of the station. The annual value of potential evapotranspiration is also termed as 'Thermal efficiency'.

Penman's Method

The formula developed by Penman for estimating P.E. has combined the turbulent transfer and energy balance approaches. This formula is:[6]

$$PE = \frac{\frac{\Delta}{\lambda}\left\{R_A(1-r)\left(a+b\frac{n}{N}\right)-6T^4(0.56-0.092\sqrt{e_d})\left(0.10+0.90\frac{n}{N}\right)\right\} + 0.35(e_a-e_d)\left(1+\frac{U}{100}\right)}{\frac{\Delta}{\lambda}+1}$$

Where, PE = Potential evapotranspiration in mm per day

R_A = Incident radiation outside the atmosphere on a horizontal surface expressed in mm of evaporable water per day

λ = Reflection coefficient or Albedo

n/N = Ratio of actual hours of sunshine to theoretical duration of sunshine

σ = Stefan - Boltzman constant

a and b constants

T = Mean temperature in °C

σT^4 = Black body radiation at mean temperature

e_a = Saturation vapour pressure in mm of mercury

e_d = Actual mean vapour pressure in mm of mercury

U = Wind speed at 2 metres above ground in miles per day

Δ = Rate of change of saturation vapour pressure with temperature in mb per °C

λ = Psychrometric constant in mb per °C

$R_A(1-r)$ (a+b n/N) represents the incoming shortwave radiation and $\sigma T^4(0.56-0.092\ \sqrt{e_d})$ (0.10+0.09 n/N) the outgoing long wave radiation; Reflection coefficient *r* (albedo) is taken as 0.25 for vegetation on the basis of Monteith's (1959) measurements. The aerodynamic term $0.35\ (1+{}^{U}/_{100})\ (e_a\text{-}e_d)$ is the drying power of the atmosphere based on saturation deficit, air movement and extra roughness of vegetation cover compared to water surface. The factor Δ/λ makes allowance for the relative significance of net radiation and aerodynamic terms.

Modifications

There modifications have been introduced in the original equation. These are:

Correction for Elevation of the Station

In the original derivation of the equation, Δ/λ is taken to depend on mean air temperature only. Recent studies have shown that this factor depends on altitude. A correction is, therefore, applied by multiplying Δ/λ by Po/Ph where *Po* is standard sea level pressure and *Ph* station level pressure.

Correction for Latitude of the Station

In the term for incoming radiation, (0.29 Cos ϕ + 0.52 n/N) is used as suggested by Glover and McCulloch (1958) for (a + bn/N), ϕ being the latitude of the station. This term takes into account the obliqueness of sun's rays traversing the atmosphere and is, therefore, considered better for studying stations distributed over wide latitudinal belts. Radiation measurements made at stations in India gave values for 'a' ranging from 0.27 to 0.38 and for 'b' from 0.33 to 0.44. Value of 'a' obtained is thus slightly greater and that of 'b' slightly smaller than given by Glover and McCulloch.

Sunshine Data

As sunshine data normal are not available for all the stations, *n*/*N* is taken uniformly for all stations as equal to (1–m), where m is the cloudiness expressed as tenths of the sky.

The formula this modified becomes:

$$PE = \frac{\left(\frac{Po}{Ph}\right)\frac{\Delta}{\lambda}[R_A(1-r)\{0.29\,\mathrm{Cos}\,\theta+0.52(1-m)-6T^4(0.56-0.092\sqrt{e_d})\ 0.10+0.90(1-m)\}] + 0.35(e_a-e_d)\left(1+\frac{U}{100}\right)}{\left(\frac{Po}{Ph}\right)-\frac{\Delta}{\lambda}+1}$$

Rowher's Method

Rowher's formula has been modified and used for estimation of potential evapotranspiration over the Indian subcontinent. It has given satisfactory results for classification of vegetation of India, Pakistan and Burma. However, his formula has not been used in the present study.

Carl Rowher's formula (1931):[7]

$$E = (1.465 - 0.0186\ B)\ (0.44 + 0.118\ W)\ (e_s - e_d)$$

This formula has been modified by Raman and Satakopan in this computation, because no data of water surface temperature is available so the term $(e_s - e_d)$ in Rowher's equation had to be approximated to $(100/h - 1)^e$ where 'h' is the relative humidity in per cent and *e* the wind velocity as recorded at different heights above the ground that vary from station to station. The height has arbitrarily been reduced to the level of 4 ft. above the ground by using a curve of Chapman giving the mean variation of wind velocity with height. Thus[8]:–

$$E = (1.465 - 0.0186\ B)\ (0.44 + 0.118\ W)\ (100/h - 1)^e$$

Where, E = Mean daily evaporation in inches during 24 hours
W = Mean daily wind velocity at the level of the evaporating water surface
B = Mean daily value of the barometric pressure in inches of mercury

This value of E should be multiplied by number of days in the month to get the monthly evaporation figure. After obtaining the monthly evaporation figure by the above equation, it is again multiplied by 0.60 to get natural evaporation from an extensive wet area with rain or irrigation. This natural evaporation from saturated ground approximates the potential evapotranspiration of Thornthwaite's terminology.

POTENTIAL EVAPOTRANSPIRATION AS A CLIMATIC FACTOR

No one can say whether a climate is moist or dry merely from the knowledge of precipitation. It is necessary to know whether the rainfall is greater or less than the water need (water required for evaporation and transpiration). The climate of a place is moist, if the precipitation exceeds the water need, but dry if it falls short of the water need, while it will be neither humid nor arid if the water need approximately equals precipitation. It will be possible to distinguish between the actual amount of water that evaporates and transpires from the optimum that would transpire and evaporate if sufficient water is available. When water supply is adequate, actual evapotranspiration rises to its maximum (equals the potential value) depending only on temperature. In addition to Thornthwaite, a comparison between potential evapotranspiration and precipitation was made by several authors like Transcau,[9] Isozaki,[10] Koppen,[11] Angstrom[12] etc. in order to get P/PE index. The values of P/PE index determine the conservative climatic parameters with respect to water need.

The other significance of potential evapotranspiration as an isolated climatic or hydrologic parameter ultimately depends upon the amount of discrepancy between P.E. and the amount of A.E. because the discrepancy between potential and actual water losses will tend to increase as the rainfall diminishes except in low-lying areas, such as river flood plains and alluvial flats, where the water table is so close to the ground surface that plants are able to draw directly by capillary action from the considerable ground water reserves.

DISTRIBUTION OF POTENTIAL EVAPOTRANSPIRATION OVER UTTAR PRADESH

Average values of monthly and annual potential evapotranspiration have been computed from available mean temperature of 39 stations in the state and computed values of 22 stations by Penman method[13] have been incorporated in the present study for comparison sake.

Potential evapotranspiration depends on the amount of solar energy received by the surface and its resulting temperature. Figure 6.1A which depicts the influence of the relative intensity of solar radiation, indicates that almost vertical rays of the sun makes the southern parts (Pipri) more vulnerable than the northern parts. The other cause is variation in length of the astronomical day, i.e., the duration of sunshine, which affects monthly potential evapotranspiration.

The two aforesaid factors, no doubt, depend on latitude, but the resultant temperatures are not controlled by latitude alone. The other obvious factors are the varying distances from the sea, altitude, monsoonal characteristics of rainfall, nature of the sky, cold wave, heat wave etc. which from a complex combination affecting the distribution of temperature and consequently the distribution of P.E. over the state.

Variations in altitude, latitude, and distance from the sea affect to a considerable extent, the distribution of potential evapotranspiration, which varies from 69.5 to 170.5 cm in the state. The isopleth of 142.5 cm runs almost E-W in the almost featureless middle parts of the Ganga plain, but in the U.P. Himalaya high elevations and diverse topography cause the computed values of P.E. to vary from 69.5 to 122.5 cm. Another striking feature is observed in the Bundelkhand area, where the influence of the Arabian sea results in lower temperature during summer, so that annual values of P.E. is lower (less than 145.0 cm) than that over the southeastern upland (170.5 cm at Pipri). Seasonal variations of P.E. at each station are of great importance from practical point of view. Table 6.1 represents the computed values of P.E. which are lowest in the month of January (less than 3 cm) except at Pipri (4.5 cm) and highest (10 to 22 cm) is in the month of May in the east U.P. and in June in the west U.P. because in the former the air temperatures are reduced in the month of June due to onset of the southwest monsoon. An estimation has been made as to how many times the P.E. in the warmest month is higher than that in the coldest month. The mean values of P.E. are about six-fold and eight-fold higher in summer as compared to those in winter in the U.P. Himalaya and in east U.P. respectively, whereas, these values in west U.P. are almost seventeen times higher in summer as compared to those in winter, because that region is affected more frequently by cold waves and western disturbances during the winter months and by heat waves from the adjacent desert in the summer months.

Monthly and annual computed values of P.E. for 22 stations by Penman method are given in Table 6.2.[14] The annual values of Penman's P.E. Fig. 6.1 are almost similar to those of Thornthwaite (Fig. 6.1A), but Penman's monthly values of P.E. are higher in winter and lower in summer than those of Thornthwaite. Therefore, the annual range of P.E. in Penman's method is not so high. In winter months of December and January the P.E. varies from 5.0 to 6.0 cm except at Dehra Dun and Roorkee where it is about 4.0 cm. It increases in the hottest months of May and June to about 20.0 cm. In Kanpur, it is still higher 25.0 cm in May and 23.0 cm in June. It is 13.0 to 14.0 cm in July and 11.0 to 12.0 cm in August (Tables 6.1 and 6.2).

TABLE 6.1: Mean Monthly and Annual Potential Evapotranspiration in cm (Thornthwaite)

Sl. No.	Stations	Jan.	Feb.	Mar.	Apr.	May	June	July	Aug.	Sept.	Oct.	Nov.	Dec.	Annual
1.	Mussoorie	1.6	1.8	4.3	7.1	10.6	10.5	9.4	8.6	7.1	5.6	3.5	2.6	72.7
2.	Dehra Dun	1.9	2.6	6.5	12.0	18.4	18.6	16.6	14.7	12.7	8.8	4.3	2.1	119.2
3.	Mukteswar	1.6	2.1	4.3	7.1	9.9	10.1	9.3	8.5	7.4	5.9	3.7	2.4	72.3
4.	Nainital	1.4	2.1	4.6	6.8	10.2	10.8	9.2	9.2	7.7	5.3	3.2	2.1	72.6
5.	Joshimath	1.4	1.8	4.0	7.1	9.9	11.6	11.2	10.3	8.3	8.2	2.9	1.9	78.6
6.	Tehri	1.6	2.4	6.2	12.0	17.7	19.0	19.4	17.1	14.2	6.8	4.3	1.8	122.5
7.	Mukhim	1.6	1.8	4.3	7.5	10.6	11.2	10.8	9.9	8.3	6.2	3.7	2.4	78.3
8.	Munsiari	1.9	2.6	3.1	6.2	7.8	10.2	10.1	8.9	7.4	5.0	3.7	2.6	69.5
9.	Roorkee	1.6	2.6	7.1	14.9	20.2	20.4	19.1	17.4	15.1	9.7	4.0	1.9	134.0
10.	Najibabad	1.6	2.6	7.3	14.9	19.8	20.3	19.4	17.4	15.5	9.7	4.3	2.1	134.9
11.	Pantnagar	1.0	2.1	7.2	13.2	18.3	19.1	18.2	17.3	14.8	9.1	3.7	1.3	125.3
12.	Bahraich	2.2	3.7	9.6	16.7	20.7	20.1	18.8	17.6	16.0	13.0	5.9	2.7	147.0
13.	Nautanwa	2.2	3.4	9.6	16.1	19.5	19.4	18.4	17.3	15.4	12.4	5.9	2.7	142.3
14.	Meerut	1.4	2.6	8.3	16.1	20.4	20.9	19.6	17.6	15.8	11.2	4.3	2.1	140.3
15.	Aligarh	1.6	3.1	8.7	16.8	21.2	21.1	20.2	18.1	16.1	12.6	5.3	2.4	147.2
16.	Agra (Obsy)	1.7	3.4	9.6	17.0	21.3	21.1	19.5	17.6	16.0	12.7	5.4	2.4	147.7
17.	Agra	1.7	3.4	9.6	17.0	21.2	21.0	19.8	17.6	16.0	12.7	5.4	2.4	147.8
18.	Mainpuri	1.9	3.4	9.6	17.0	20.9	21.1	19.8	18.0	16.0	13.3	5.9	3.8	150.7

Contd.

TABLE 6.1 *Contd.*

Sl. No.	Stations	Jan.	Feb.	Mar.	Apr.	May	June	July	Aug.	Sept.	Oct.	Nov.	Dec.	Annual
19.	Jhansi	2.8	4.5	12.1	17.9	21.1	20.9	18.6	16.8	15.3	13.1	6.3	3.3	152.7
20.	Orai	2.2	4.2	11.4	17.7	21.2	21.0	19.0	16.8	15.6	13.1	5.7	2.7	150.6
21.	Banda	2.5	4.5	12.7	17.9	21.1	20.9	19.0	18.1	15.9	13.4	6.0	3.3	155.3
22.	Bareilly	1.9	3.4	9.3	16.7	20.7	20.5	18.2	17.6	16.1	12.3	5.4	2.4	144.5
23.	Kheri	1.9	3.7	9.6	16.7	20.7	20.2	19.5	18.3	15.8	12.3	5.7	2.7	147.1
24.	Hardoi	1.9	3.7	14.2	16.7	20.6	20.8	19.1	17.6	16.0	12.4	5.4	2.7	151.1
25.	Lucknow (Obsy)	2.5	4.0	10.5	17.3	20.9	20.7	19.1	17.6	15.9	12.8	5.4	2.7	149.4
26.	Lucknow	2.5	4.0	10.5	17.3	20.5	20.4	18.8	17.6	15.9	12.8	5.9	2.7	148.9
27.	Kanpur (Obsy)	2.2	3.7	10.5	17.3	21.2	20.7	19.0	17.5	15.6	12.8	5.4	2.4	148.3
28.	Kanpur	2.2	3.7	10.5	17.3	21.2	20.7	19.0	17.5	15.6	12.8	5.4	2.5	148.4
29.	Fatehpur	2.5	4.0	10.8	17.7	21.2	21.0	19.3	17.5	15.9	13.1	6.0	2.7	151.7
30.	Allahabad	2.5	4.3	11.7	17.5	21.1	20.9	19.0	17.6	15.9	13.4	6.0	2.7	152.6
31.	Gonda	2.2	3.4	9.6	16.7	20.5	20.0	18.8	17.6	15.6	12.5	5.4	2.4	144.7
32.	Gorakhpur	2.5	4.0	11.4	17.3	10.2	19.7	18.7	17.6	15.9	13.7	6.2	3.3	150.5
33.	Azamgarh	2.5	4.2	11.4	17.7	20.9	20.4	18.6	17.5	15.6	13.7	6.3	3.3	152.1
34.	Varanasi (Obsy)	2.5	4.3	11.7	17.5	20.8	20.6	19.0	17.5	15.9	13.4	6.0	2.7	151.9
35.	Varanasi	2.5	4.3	11.7	17.5	20.8	20.6	19.0	17.5	15.9	13.4	6.0	2.7	151.9
36.	Faizabad	2.0	4.0	9.9	17.7	19.5	19.7	18.7	17.6	15.3	11.6	5.4	2.4	143.8
37.	Sultanpur	2.2	4.0	10.8	17.3	20.5	20.4	18.7	17.5	15.9	12.8	6.4	2.4	148.9
38.	Ballia	2.8	3.7	9.9	15.7	20.8	20.3	19.0	17.5	15.9	13.4	4.9	3.3	147.2
39.	Pipri	4.5	7.7	15.1	18.4	21.0	20.5	18.8	17.5	16.2	15.7	9.6	5.5	170.5

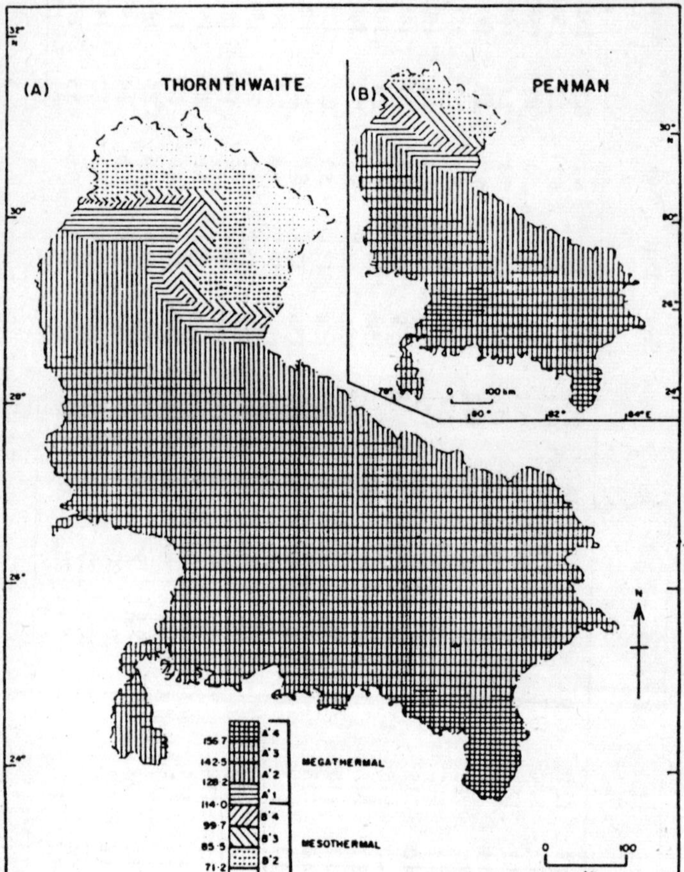

FIG. 6.1: Average Annual Potential Evapotranspiration: (A) Thornthwaite, (B) Penmann

A comparative study of these two methods reveals the fact that the values of P.E. computed by Penman's method are higher in winter than those computed by Thornthwaite's method but the relationship is *vice versa* in the summer months. Therefore, Penman's formula is

TABLE 6.2: Mean Monthly and Annual Potential Evapotranspiration in cm (Penman)

Sl. No.	Stations	Jan.	Feb.	Mar.	Apr.	May	June	July	Aug.	Sept.	Oct.	Nov.	Dec.	Annual
1.	Dehra Dun	3.79	5.56	9.66	13.42	16.89	16.21	11.23	10.23	10.35	9.15	5.22	3.53	115.24
2.	Roorkee	4.28	6.24	10.04	15.27	19.89	19.20	13.35	12.35	12.16	9.94	5.55	3.85	133.33
3.	Bahraich	5.11	7.34	12.92	17.64	21.26	18.92	13.24	11.79	12.08	11.46	6.95	4.78	143.49
4.	Meerut	5.31	7.51	12.71	17.47	22.22	22.53	15.30	14.21	14.22	11.13	6.59	4.33	154.90
5.	Aligarh	5.38	7.62	12.98	17.45	22.18	21.47	15.11	12.71	13.63	11.94	7.29	5.15	152.91
6.	Agra	5.29	7.52	12.64	16.26	20.83	20.44	14.58	12.39	13.29	11.61	6.87	4.97	146.69
7.	Mainpuri	5.02	7.25	12.51	16.50	20.91	19.48	13.58	12.11	12.63	11.49	6.87	4.77	143.12
8.	Jhansi	6.35	8.49	13.83	17.23	21.51	20.11	13.65	11.63	12.70	12.28	7.91	5.86	151.55
9.	Orai	5.86	8.53	14.72	18.80	23.87	22.07	14.64	12.20	13.37	12.74	7.93	5.53	160.26
10.	Banda	5.44	7.79	13.02	16.72	21.03	19.51	13.68	12.10	12.80	11.50	7.02	5.01	145.62
11.	Bareilly	4.88	7.06	12.47	16.65	20.85	19.60	12.94	11.55	12.26	11.01	6.42	4.53	140.22
12.	Kheri	4.67	7.12	12.09	16.75	19.17	18.19	12.80	11.98	12.32	10.93	6.51	4.38	136.91
13.	Hardoi	6.03	7.77	13.13	17.49	21.47	20.00	14.18	12.68	13.33	11.42	6.95	4.88	149.33
14.	Lucknow (Obsy.)	6.68	8.20	14.17	18.99	22.97	20.95	13.15	11.35	12.16	11.35	7.07	4.86	151.90
15.	Lucknow	5.16	7.34	12.55	16.43	18.53	18.37	12.57	11.64	12.28	11.36	6.88	4.75	137.86
16.	Kanpur	6.07	8.88	15.73	20.53	25.38	22.98	14.66	12.54	13.28	12.55	7.83	5.60	166.03
17.	Fatehpur	5.54	7.77	13.55	17.36	21.02	19.60	13.63	11.72	12.28	11.60	7.23	5.11	146.41
18.	Allahabad	6.19	8.60	14.71	18.78	23.34	20.58	12.77	11.03	12.01	12.16	7.82	5.70	153.69
19.	Gonda	5.18	7.48	13.46	17.79	20.65	18.33	12.84	11.65	11.95	11.28	6.88	4.74	142.23
20.	Gorakhpur	5.35	7.65	13.55	17.72	20.72	17.11	13.06	11.93	11.61	11.47	7.18	5.02	142.37
21.	Varanasi (Obsy.)	6.11	9.55	15.73	21.03	24.91	20.44	13.19	11.67	12.13	12.32	8.01	5.73	160.82
22.	Varanasi	6.37	8.87	14.99	18.82	22.91	18.55	12.71	11.38	11.86	12.05	8.05	5.95	152.46

more suitable for the tropical and sub-tropical zones where seasonal differences in the computed values of P.E. are not so great. This observation is based on the results obtained by applying the two methods in determining P.E. values for some selected stations in the study area.

NOTES

1. Shanbhag, G.V.: "Some Notes on Evapotranspiration, Evaporation from Soil and Transpiration", *Ind. Jour. Met. and Geophy.*, Vol. 8, No.2, New Delhi (1957), pp. 127-136.
2. Thornthwaite, C.W.: op cit. (1948), pp. 55-94.
3. Penman, H.L.: The Physical Basis of Irrigation Control, Proc. Int. Horticultural Congress, London (1952), pp. 913-24.
4. Chow, V.T. (Ed.): *Evapotranspiration, Handbook of Applied Hydrology*, McGraw-Hill Book Company, New York (1964), Section II, pp. 1-38.
5. Thornthwaite, C.W. and Mather, J.R.: op. cit,., Vol. 10, No. 3 (1957).
6. Rao, K.N. et al.: Potential Evapotranspiration (P.E.) Over India, *Scientific Report* No. 136, I.M.D. (1971).
7. Rowher, C.: Evapotranspiration from Free Water Surface, Technical Bulletin No. 271, Deptt. of Agriculture, Washington (1931).
8. Ramdas, L.A.: Evapotranspiration and Potential Evapotranspiration Over the Indian Subcontinent, *Indian Journal of Agricultural Science*, Vol. 27 (1957), pp. 137-149.
9. Transcau, E.N.: Forest Centres of Enmerica, *Amer. Nat.*, Vol. 39, pp. 875-889.
10. Isozaki, M.: Thornthwaite's New Classification of Climate and its Application to the Climate of Japan, *Journal of Geography*, Vol. 45, Tokyo, pp. 234-245 (1946).
11. Koppen, W.G.: Handbuch der Klimatologic, *Gebruder Borntragger*, Vol.1, Berlin (1931-36), quoted by Subrahmanyam.
12. Angstrom, A.: A Coefficient of Humidity of General Applicability, *Geografiska Annular*, Vol. 18, Stockholm Sweden (1936), pp. 245-254.
13. Rao, K.N. et al.: op cit., No. 136 (1971).
14. Rao, K.N. et al., *Ibid*, Table 1, pp. 2-3.

7

COMPUTATION OF WATER BALANCE AND ITS DIFFERENT COMPONENTS

CLIMATIC WATER balance is one of the basic problems in modern climatology, but it has also been the centre of discussion in classical climatology. The state of water balance in any region represents its climatic conditions, which determine the climatic parameters of a region as conceived by modern thinkers and researchers in this field. Hence, it involves the study of water balance as well as of heat balance, in both of which it is necessary to determine the rate and amount of solar radiation. Water balance refers to the balance between the income of water from precipitation and outflow of water by way of evapotranspiration, surface run-off and infiltration to the soil. Heat balance is dependent upon the solar radiation and its disposal back to the outer space, whereas water balance involves both the factors like solar radiation and precipitation, the former constituting the source of soil temperature and the latter the source of soil moisture.

The term 'water balance' has recently gained widespread popularity among Climatologists, Geographers, Geologists, Geophysicists and others concerned with water problem. Its study was advanced significantly by the works of Thornthwaite (1943) in the United States, Penman (1948) in England and Budyko (1956) in the U.S.S.R. The present study has been made about water balance of Uttar Pradesh on the basis of Thornthwaite's 1955 concept.

In order to compute the monthly water balance at a place it is necessary to have the following scientific informations:

(i) Mean monthly temperature
(ii) Mean monthly precipitation
(iii) Necessary conversion and computational tables
(iv) Information on the water holding capacity of the depth of soil for which the balance is to be computed.

First two items are easily available at all the meteorological stations where rainfall and temperature are measured, and the third is obtained from the report of Thornthwaite and Mather.[1] It is comparatively difficult to obtain information regarding the water holding capacity of the soil, which depends on type and structure of the soil as well as on the type of vegetation covering the surface. A sandy soil will hold only a cm or two of moisture per 30 cm depth of soil while silt or clay may hold ten cm or more of moisture in the same depth. Water holding capacity of different soils under different vegetal cover is given in the Thornthwaite's book in Table 10.[2] In 1948, Thornthwaite assumed 10 cm as the soil moisture holding capacity but as a result of latter work in 1955, in an attempt to provide more realistic approach, the soil moisture holding capacity was taken to be 30 cm on an average for the world. It varies from 2.5 cm to 40.0 cm depending upon the nature of surface. Prof. V.P. Subrahmanyam has suggested the nearest possible vegetation types and water holding capacity on the basis of the soil characteristics and their types in the state as follows (Table 7.1).

TABLE 7.1: Soil Types and Their Moisture Holding Capacity

Soil types[3]	Natural vegetation	Moisture holding capacity (cm)
Brown hill soils (on sandstone and shales)	Temperate forest type vegetation	5.0
Tarai soils	Tall grasses	15.0
Alluvial soils	Shrubs and low grasses	25.0
Calcarious soils	Shrubs and low surface vegetation	15.0
Medium black soils	Moderately deep rooted vegetation	20.0
Mixed red and black soils	Shrubs and pastures	20.0
Red sandy soils	Pasture and shrubs	10.0
Red and yellow soils	Forest type vegetation	40.0

The two terms 'field capacity' and 'moisture holding capacity' though resemble are not the same. The moisture holding capacity is that, when all pores of the soil are filled with water which is referred to as saturation and occurs after prolong rains before the gravitational 'free water' has time to be drained out. The field capacity is the maximum amount of water that can be retained in the soil against gravity. When the moisture content goes below a certain point then water cannot be withdrawn from the soil by plants. This is defined as 'wilting point'. The water lying between the field capacity and the wilting point is known as "available water'. This quantity is utilised for actual evapotranspiration when the precipitation falls short of P.E.

WATER BALANCE COMPONENTS

The important components of water balance are:

1. Potential evapotranspiration
2. Precipitation
3. Actual evapotranspiration
4. Soil moisture utilisation
5. Soil moisture recharge
6. Water deficit
7. Water surplus.

Potential evapotranspiration and precipitation are the important factors, because with the help of these two, the other water balance components may be obtained. When the potential evapotranspiration is compared with precipitation it is called as precipitation effectiveness (Thornthwaite's monthly values of precipitation effectiveness index was derived from monthly values of precipitation and temperature). A negative value of P-PE indicates the amount by which the precipitation falls short of the potential evapotranspiration of an area. A positive value, on the other hand, indicates the amount of excess water which is available, during a certain period or periods of the year, for soil moisture recharge and run-off. The state as a whole enjoys alternate wet and dry seasons, i.e., the monthly precipitation is less in one season and more in the other than the water need and hence, corresponding to that condition moisture deficit or surplus would occur. Water balance computation for different stations in the state has revealed single as well as multiple wet and dry seasons. At a particular station if precipitation is always more than the P.E. the soil will remain full of water and surplus 'S' will occur. At another place if in every month precipitation falls short of the potential evapotranspiration, there will not be enough moisture for vegetation to use and a moisture deficit 'D' will occur. However, such stations are conspicuous by their absence in U.P. At many stations period of moisture surplus are followed by period of deficient precipitation during which the stores soil moisture is used, even then the actual evapotranspiration falls short of the potential and a moisture deficit 'D' occurs. Thus, stations with both wet and dry seasons as well as with cold season of low water need normally show[4] (a) a period of full storage, when precipitation exceeds the water need and a moisture surplus 'S' results, (b) the early part of a dry period, when the stored soil moisture and rainfall are used for actual evapotranspiration, is termed as period of soil moisture utilisation, (c) when soil moisture storage is steadily diminished, the actual evapotranspiration falls much below the potential and a moisture deficit 'D' occurs, (d) a moist season when precipitation again starts exceeding the water need, soil moisture is recharged till it attains its full capacity, and (e) after saturation of the soil precipitation exceeding the water need is counted as water surplus 'S'. Thornthwaite and Mather have assumed in their book that about 50 per cent of surplus water will run-off in a given month and the rest is retained for run-off in the subsequent month signifying the importance of time lag.[5] However, in any region the rate of removal by run-off depends on the size of the watershed, nature of its slope, ground cover and the type of soil through which water has to move. The water balance equation is expressed as:

$$\text{Precipitation} = \text{Potential evapotranspiration} - \text{Deficit} + \text{Surplus} \pm \text{Storage change}$$

EVALUATION OF WATER BALANCE IN UTTAR PRADESH

An attempt has been made to compute water balance of 39 stations, where recorded monthly rainfall and temperature data are available. Water balance of 27 selected places are illustrated in the graphs, which depict the nature and variation of monthly water balance components. These are Dehra Dun, Mussoorie, Nainital (Fig. 7.1); Nautanwa, Pantnagar, Najibabad, Tehri, Mukteswar, Joshimath (Fig. 7.2); Pipri, Jhansi, Kanpur, Agra, Meerut, Roorkee (Fig. 7.3); Allahabad, Varanasi, Gorakhpur, Gonda, Bareilly, Ballia (Fig. 7.4); Bahraich, Kheri, Aligarh, Mainpuri, Banda and Lucknow (Fig. 7.5).

Water balance computation provides a good insight with respect to the moisture position of an area. The potential evapotranspiration is less in winter but in early summer is begins to rise rapidly reaching the highest values in May or June at almost all stations of the State. It falls slowly during the rainy months. The corresponding precipitation is unevenly distributed throughout the year (Fig. 3.5). The rainy months are usually later part of June, July, August and September, which receive 120 to 200 cm in the U.P. Himalaya, 80 to 120 cm in the east U.P. and 60 to 80 cm in the west U.P. On the other hand, less than 40 cm of rain in the U.P. Himalaya and about 15 cm of rain in the rest of the state are received in other seasons of the year.

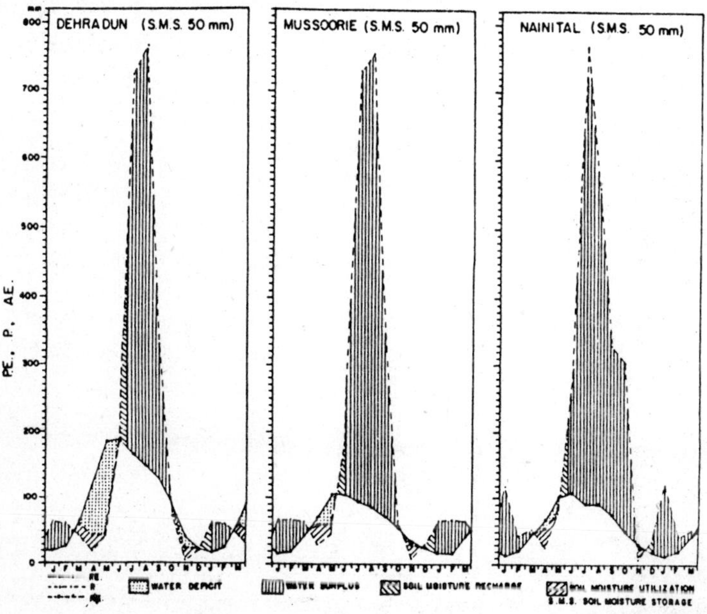

FIG. 7.1: Thornthwaite Water Balance Graph for Dehra Dun, Mussoorie, Nainital

It is needless to emphasise that water need depends on temperature and length of days. We also know that rainfall and water need do not vary directly. At the same time of maximum rainfall in July, August and September, there is more water need because of higher temperature rather than in December, January and February because of cold weather. Likewise, in the U.P. Himalayan region temperature decreases with increasing altitude and potential evapotranspiration is also less. Therefore, the periods of soil moisture utilisation as well as that of water deficit, to some extent, occur during March-May and October-November at places like Dehradun, Mussoorie, Mukteswar and Nainital which record a water deficit of 24.3, 15.1, 3.5 and 6.3 cm respectively. During the early period of rainy season the water need is less than precipitation and the surplus rain first rechanges soil moisture, that was used up previously and then raises the ground water level and produces surface and subsurface runoff when the soil becomes saturated. Thus, the soil moisture recharge and water surplus periods occur from June to September and December to February. The average annual water surplus for the aforesaid stations has been computed as 136.5, 170.6, 68.8 and 190.7 cm respectively, which is due to heavy orographic rainfall during the southwest monsoon season and by the western disturbances in winter. When precipitation is less than potential evapotranspiration, the difference is made up in part from the soil moisture storage, but as the soil becomes drier the part not made up becomes increasingly larger.

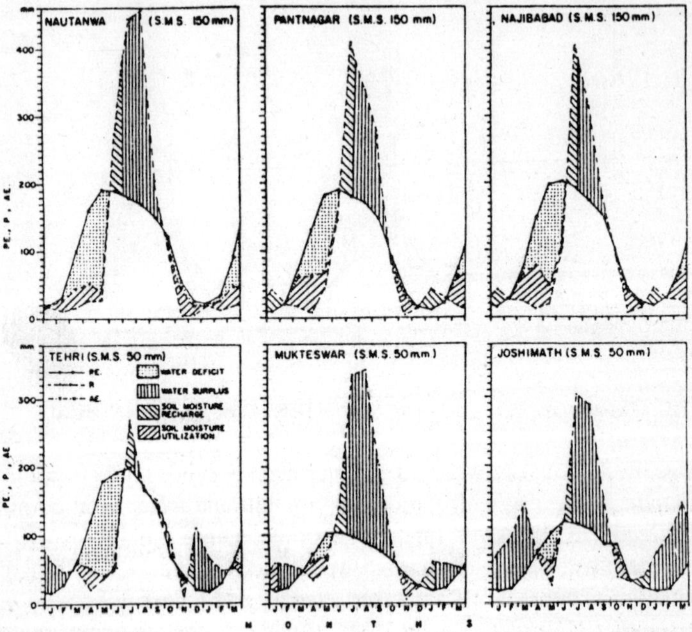

FIG. 7.2: Thornthwaite's Water Balance Graph for Nautanwa, Pantnagar, Najibabad, Tehri, Mukteswar, Joshimath

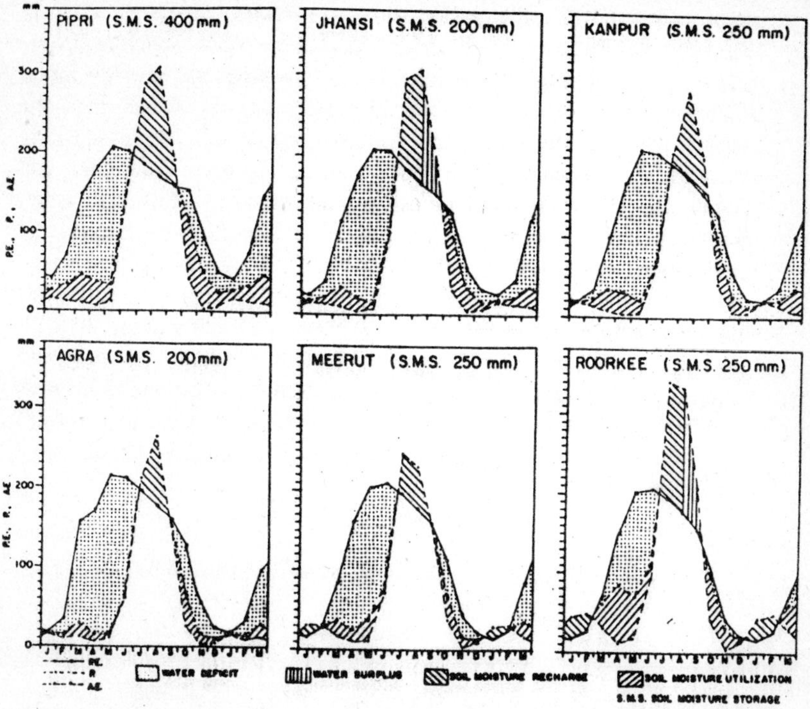

FIG. 7.3: Thornthwaite's Water Balance Graph for Pipri, Jhansi, Kanpur, Agra, Meerut, Roorkee

Computation of water balance for Ballia, Pipri and western parts of the state like, Agra, Aligarh, Meerut, Mainpuri, Kanpur and Bareilly (Semiarid) do not show any water surplus although the water need is less than precipitation during June to September because the surplus rain is totally spent in soil moisture recharge. The average annual water deficit is greater at these places, the amount being 49.8, 64.7, 79.8, 69.8, 57.5, 74.2, 60.1 and 57.5 cm respectively. These figures are comparatively higher than in other parts of the State. Jhansi and Banda also come under Semiarid type of climate, the water deficit at these places being 66.2 and 73.7 cm respectively, but they reveal some surplus in the months of August and September (9.4 and 18.9 cm).

The water balance of Pantnagar and Nautanwa represent the climatic conditions in the Tarai region (Moist subhumid) extending as a narrow belt from west to east along the Himalayan foothills. In this region soil moisture recharge also occurs in January and February in winter and again starts in June during summer. The water surplus has been computed for the period July to October as 38.1, 53.6 cm respectively. The average water deficit at these places amount to 23.3 and 34.7 cm during the periods March to June and in November. Comparatively speaking the western parts are more wet than the eastern parts of the Tarai. The eastern parts of U.P. represented by Varanasi, Allahabad, Bahraich, Gonda, Kheri, Lucknow and Gorakhpur come under Dry subhumid climate. Therefore, water surplus occurs only in the wet and warm months of August and September, which ranges between 0.2 and 25 cm. It has been found more wet than the western Semiarid parts during the southwest monsoon season. However, great variation in the net amount of P-PE are observed in the State. It varies in the U.P. Himalaya and the Tarai regions from 187.2 to –1.9 cm. In the western Semiarid region it varies from –17.6 cm at Roorkee to –79.8 cm at Agra whereas in the eastern Dry subhumid region the variation is from –15.3 cm at Gonda to –50.2 cm at Lucknow.

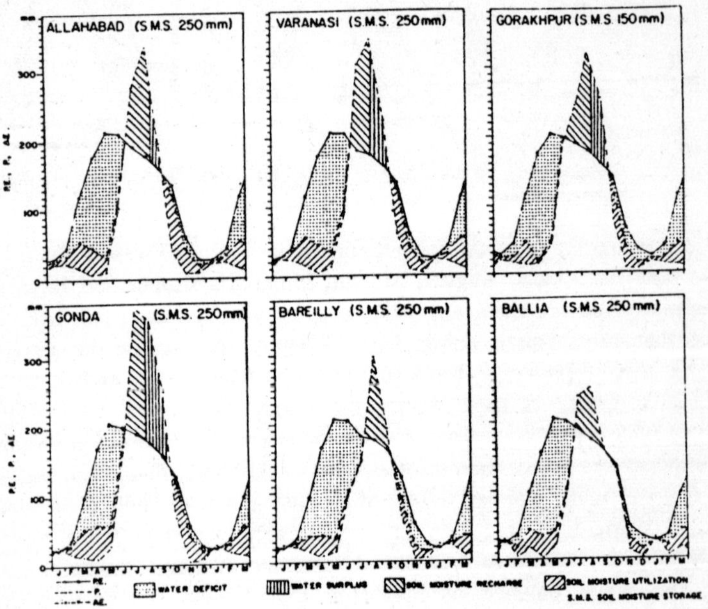

FIG. 7.4: Thornthwaite's Water Balance Graph for Allahabad, Varanasi, Gorakhpur, Gonda, Bareilly, Ballia

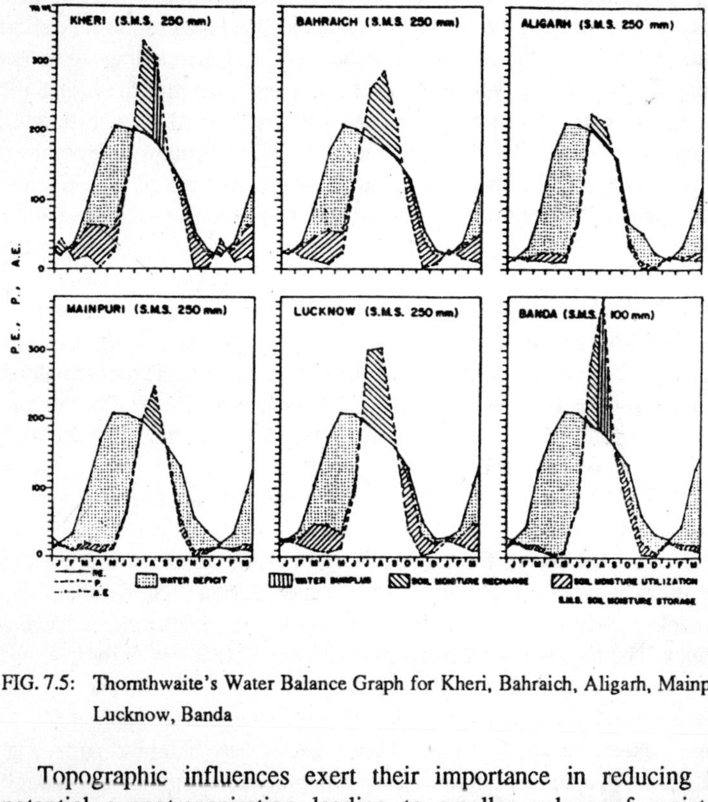

FIG. 7.5: Thornthwaite's Water Balance Graph for Kheri, Bahraich, Aligarh, Mainpuri, Lucknow, Banda

Topographic influences exert their importance in reducing the potential evapotranspiration leading to smaller values of moisture deficit and greater values of moisture surplus because of increased orographic precipitation and lower temperature due to high elevations. In the U.P. Himalaya considerable variation in all water balance components are observed, which may be ascribed to the effects of varying altitudes causing considerable differences between monthly and annual precipitation and potential evapotranspiration. The net variation in this region is comparatively higher than in other parts of the State.

AVERAGE ANNUAL WATER SURPLUS AND DEFICIT

Figure 7.6 reveals the average annual water surplus and deficit of the State. Water deficit is computed by subtracting actual evapotranspiration from potential evapotranspiration. Even after utilising a part of soil moisture, the actual evapotranspiration falls short of the potential in dry season and the soil moisture storage goes on diminishing steadily. Whereas, water surplus is computed during the wet season when precipitation exceeds the water need after recharging the soil moisture to its field capacity. The U.P. Himalayan area has a considerable surplus (from 33.9 cm to 199.3 cm) which provides the head water of the prominent river systems of the Yamuna, the Ganga and the Ghaghra. Mussoorie, Nainital and Munsiari for instance, have surplus of 170.6, 190.7 and 199.3 cm. However, paucity of climatological stations make it difficult to study regional variations in this area. Enormous water surplus of this area follows a drought season of at least two months duration (April and May). The Tarai area along with the submontane strip has a marked water surplus of 25 to 50 cm. It is noteworthy that in central and eastern parts of the State, surplus does not occur until a month elapses after the onset of the southeast monsoon. In these areas, the computed water surplus comes up to 25 cm (Fig. 7.6A).

Figure 7.6B shows the distribution of average annual water deficit. The most prominent feature of the water deficit map is the large amount of deficit (50 to 80 cm) in those areas, which have no water surplus. The region with a considerable water deficit lies in the western part. Mirzapur and Ballia, though they lie in the eastern part of the state, fall under this category. There are also large water deficit (40 to 60 cm) extending into the central and eastern Dry subhumid parts. The U.P. Himalaya and Tarai areas have a deficit of 0.3 to 34.7 cm though the water need is comparatively meagre. Thus, the water deficit map reveals the precarious nature of water balance in many areas of the state. The impact of these deficiencies on the local economy is rendered more serious by the excessive variability of the monsoon rain. Consequently, the use of mean monthly data in an estimation of water surplus and deficit on annual basis, does not reveal a clear perspective of the water problem of the state to its full extent.

MOISTURE INDICES

Studies have shown that moisture indices can be used to differentiate the different climates found in the State. These indices are:

1. Humidity index (I_h)

2. Aridity index (I_a)
3. Moisture index (I_m)
4. Moisture adequacy index (I_{ma})

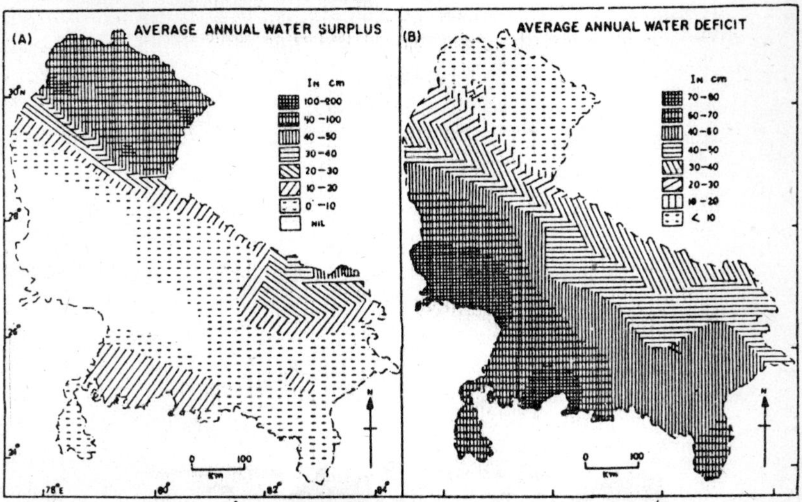

FIG. 7.6: (A) Average Annual Water Surplus, (B) Average Annual Water Deficit

Thornthwaite (1952) and Mather (1959) have shown that the vegetation association is related to the two factors which make up the moisture index, namely, the Aridity index I_a = 100 (D/PE) and Humidity index I_n = 100 (S/PE). They have ranked vegetation classes correlating them with decreasing Humidity index and increasing Aridity index as far as possible. The average values of Humidity and Aridity indices have been obtained for the State and depicted on Figs. 7.7A and 7.7B. The ultimate limit in the scale of aridity arises when there is no precipitation and water deficit is consequently equal to the water need making I_a 100 per cent. As precipitation and potential evapotranspiration are independent of each other, the index of humidity cannot have any upper limit. Humidity and Aridity indices are similar to water surplus and water deficit. Therefore, distribution of these indices on the maps follow almost the same trend as those of water surplus and deficit.

The Moisture index, obtained by comparing the water need at a place with the difference of moisture surplus and deficit, is an essential tool for classification of climate. It is used to delimit nine separate moisture regions. Thornthwaite's expression for the Moisture index is

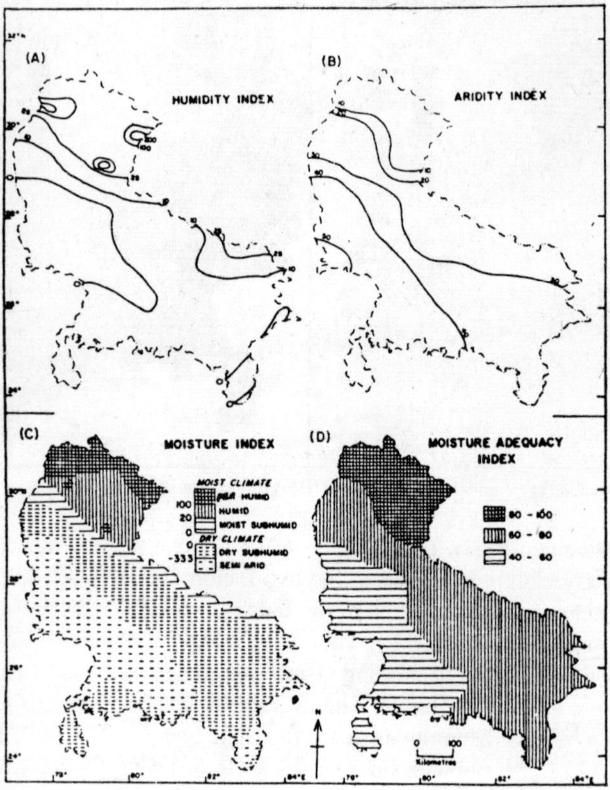

FIG. 7.7: Moisture Indices: (A) Humidity Index, (B) Aridity Index, (C) Moisture Index, (D) Moisture Adequacy Index

$I_m = I_h - I_a$ or 100(S-D)PE.[6] The calculated Moisture index of different stations in the state are shown in Fig. 7.7C. This index varies from above 100 to below –33.3, i.e., Perhumid to Semiarid in the State.

Perhumid climate occurs in those places where the elevation is above 2000 m. Other parts of the U.P. Himalaya, where the altitude is below 2000 m, come under Humid climate. The Tarai belt comes under Moist subhumid climate (I_m = 0 to 20). Eastern central and southern parts are typically Dry subhumid (I_m = 0 to −33.3). Moisture index is below −33.3 in the western districts of Meerut, Aligarh, Agra, Mainpuri, Kanpur and Fatehpur as well as in Mirzapur and Ballia in east U.P., signifying Semiarid climate. Distribution of all the above climatic types are clearly visualised from Fig. 7.7C on which the relevant isopleths have been interpolated.

The term 'Moisture adequacy index' was used by Subrahmanyam and others. It indicates moisture availability in relation to water need or, in other words, the agricultural suitability of a region. They define it in terms of actual evapotranspiration expressed as percentage of potential evapotranspiration [(I_{ma} = 100 (AE/PE)].[7] Moisture adequacy index has been calculated for all the climatological stations and the results have been mapped (Fig. 7.7D). Values exceeding 80 per cent are found in the U.P. Himalaya which is typified by forest vegetation. In the Tarai region and east U.P. the I_{ma} varies from 60 to 80 per cent, the area being suitable for rice cultivation. The region of wheat, Jowar and Bajra cultivation in west U.P. is roughly bounded by I_{ma} isopleths of 40 and 60 per cent. Where the natural conditions are unsuitable for growing Wheat the I_{ma} values lie below 40 per cent at Agra and its surroundings. The concept of Moisture adequacy index is of immense importance and it can be applied to assess the suitability of an area for certain crops and vegetations.

A COMPARATIVE STUDY OF WATER BALANCE BASED ON PENMAN'S[8] AND THORNTHWAITE'S METHODS OF COMPUTING POTENTIAL EVAPOTRANSPIRATION

An attempt has been made to compare the water balance based on Thornthwaite's method with that of Penman. The selected stations are Dehra Dun, Roorkee, Gonda, Jhansi, Meerut (Fig. 7.8); Kheri, Bahraich, Aligarh, Mainpuri, Banda, Lucknow (Fig. 7.9); Kanpur, Allahabad, Varanasi, Gorakhpur, Agra and Bareilly (Fig. 7.10). A comparative study of the results based on these two methods reveals the fact that the computed values of potential evapotranspiration by Penman's method are higher in winter than those obtained by

Thornthwaite's method in the whole of the State. This relationship, however, gets reversed in the summer months. In general, potential evapotranspiration increases sharply from April to May in both the concepts, but after June, Penman's potential evapotranspiration curve declines assuming a concave shape till the month of September, whereas, the Thornthwaite's potential evapotranspiration curve, during that period, assumes a convex form while descending. After the withdrawal of the monsoon, potential evapotranspiration shows a slight rise due to clearing of the sky and then decreases gradually due to low air temperature. Since, in this study area, about 70 to 90 per cent of the rain occurs during the southwest monsoon (June to September), potential evapotranspiration differences, based on the two methods, cause conspicuous differences in the computed values of water surplus and water deficit during this season as well as in the Moisture indices. Even at places like Agra, Kanpur, Bareilly and Bahraich, where Thornthwaite's computational method reveals no water surplus because the positive P-PE balance is completely utilised in soil moisture recharge, the application of Penman's method shows a water surplus of 2.44, 4.42, 0.85 and 14.69 cm respectively. According to Penman's method comparatively more water surplus occurs at places like Roorkee (25.83 cm), Gonda (37.36 cm), Jhansi (21.92 cm), Banda (31.62 cm), Lucknow (17.24 cm), Allahabad (19.99 cm), Varanasi (25.91 cm) and Gorakhpur (30.99 cm) than those obtained by Thornthwaite's method, viz., 13.7, 22.7, 9.4, 18.9, 1.9, 3.9, 12.2 and 14.0 cm respectively. Likewise slightly higher annual water deficit occurs at all the selected stations of the state due to higher estimation of potential evapotranspiration from December to February in Penman's method as compared to that by Thornthwaite. It is evident from Table 7.2.

As moisture index computation is based on water surplus and water deficit, the differences of Moisture index have also been obtained, but these differences are not significant. Hence, on the basis of both the methods, Jhansi, Meerut, Aligarh, Mainpuri, Kanpur, Agra and Bareilly come under Semiarid climate. Slight differences from Semiarid to Dry subhumid have been marked at Banda and Lucknow. Other representative stations, i.e., Roorkee, Gonda, Bahraich, Kheri, Allahabad, Varanasi and Gorakhpur come under Dry subhumid according to both the concepts. Dehra Dun is the only representative station of U.P. Himalaya, which comes under Perhumid climate. Lack of meteorological stations hampered a thorough comparison of the two methods as applied to this region. The conspicuous feature of water balance based on Penman's P.E. values is, that the water deficits during the southwest monsoon period are lower but during the winter months of December, January and February they are higher in comparison to Thornthwaite's results. Thus, Penman's results are quite close to reality as perceived in these seasons. The present authoress, however, feels that monthly potential evapotranspiration data computed by Penman's method be made available for a large number of stations and for a longer span of time.

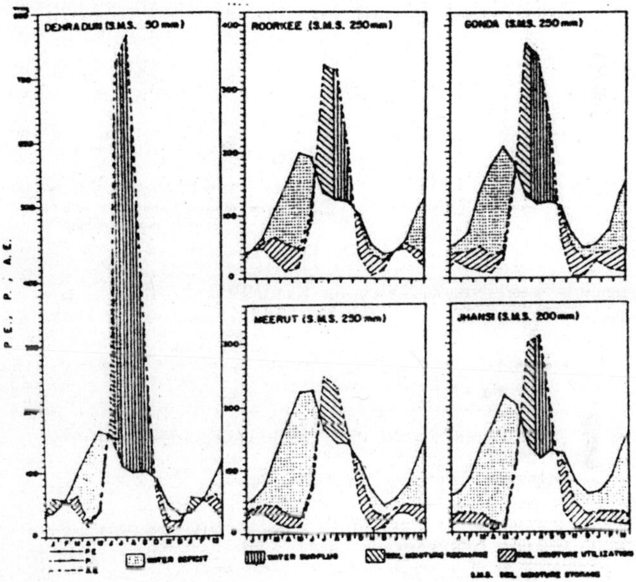

FIG. 7.8: Penman's Water Balance Graph for Dehra Dun, Roorkee, Gonda, Meerut, Jhansi

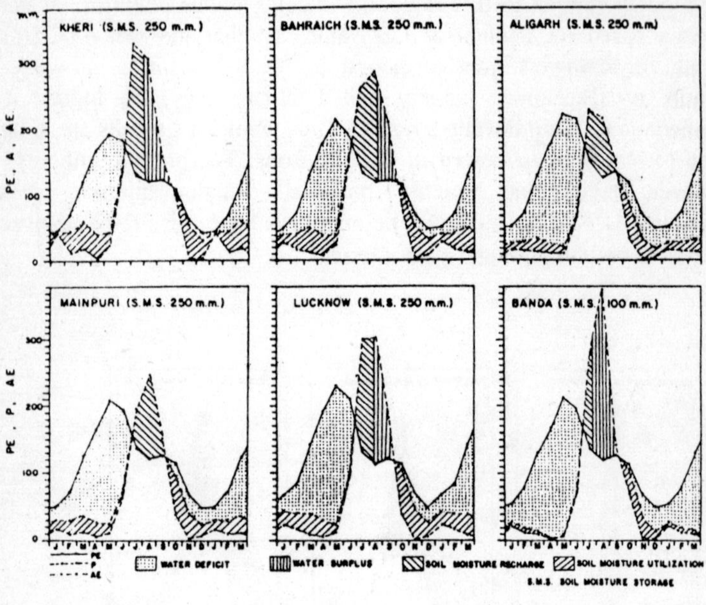

FIG. 7.9: Penman's Water Balance Graph for Kheri, Bahraich, Aligarh, Mainpuri, Lucknow, Banda

TABLE 7.2: Water Surplus, Water Deficit and Moisture Index on the Basis of Thornthwaite's and Penman's Water Balance

Stations	Water Surplus (cm)		Water Deficit (cm)		Moisture Index	
	Thornth-waite	Penman	Thornth-waite	Penman	Thornth-waite	Penman
Dehra Dun	136.5	148.88	24.3	30.72	94.13	100.80
Roorkee	13.7	25.83	31.3	42.76	−13.14	−17.70
Bahraich	0.2	14.69	39.5	50.48	−26.73	−24.94
Meerut	0.0	0.00	57.5	70.73	−40.98	−45.77
Aligarh	0.0	0.00	69.8	75.51	−47.42	−49.38
Agra	0.0	2.44	71.3	72.53	−48.21	−47.78
Mainpuri	0.0	0.00	74.2	66.92	−49.34	−46.76
Jhansi	9.4	21.92	66.2	73.57	−34.57	−34.09
Banda	18.9	31.62	73.2	76.74	−35.29	−30.99
Bareilly	0.0	0.85	57.5	53.97	−39.79	−37.88
Kheri	3.0	18.80	36.9	42.51	−23.04	−17.32
Lucknow	1.9	17.24	52.7	69.94	−33.20	−34.69
Kanpur	0.0	4.42	60.1	82.25	−40.53	−46.88
Allahabad	3.9	19.99	53.8	70.98	−32.70	−33.17
Gonda	22.7	37.36	38.0	50.19	−10.57	−9.02
Gorakhpur	14.0	30.99	47.1	55.96	−22.00	−17.54
Varanasi	39.36	25.91	12.2	75.33	−26.18	−30.73

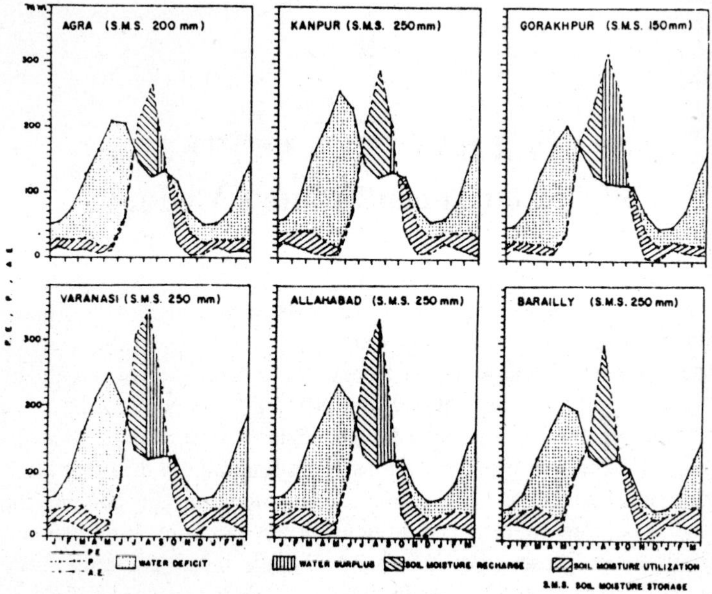

FIG. 7.10: Penman's Water Balance Graph for Agra, Kanpur, Gorakhpur, Varanasi, Allahabad, Bareilly

NOTES

1. Thornthwaite, C.W. and Mather, J.R.: Instructions and Tables for Computing Potential Evapotranspiration and Water Balance, *Publication in Climatology*, Centerton (New Jersey), Vol.10, 3 (1957).
2. Thornthwaite, C.W. and Mather, J.R.: *Ibid.*
3. Tewari, P.S.: *Agriculture Atlas of Uttar Pradesh*, Plate 12, Pantnagar (1973).
4. C.W. Thornthwaite and Mather, J.R.: 'The Water Balance, *Publication in Climatology*, Centerton (New Jersey), Vol. 8(i) (1955), p. 23.
5. Thornthwaite, C.W. and Mather, J.R.: op. cit., Vol.10, 3 (1957).
6. Mather, J.R.: op. cit. (1975), p.114.
7. Subrahmanyam, V.P. et al.: Moisture Adequacy in Relation to the Distribution of Some Crops in India, I.A.S.H. Gen. Ass. Berkeley cttee for Evaporation, Publication 62 (1963), p. 301.
8. Rao, K.N. et al.: op. cit. (1971).

8

WATER BALANCE PATTERNS (Based on Thornthwaite's Method)

WATER BALANCE is computed on the basis of two important climatic elements, because their variations are more significant in the study of water problem. In any region, temperature depends upon latitude, altitude, degree of continentality and properties of the atmosphere. It also varies from year to year due to deviations from normal in the global circulation of the upper atmosphere. Rainfall, on the other hand, varies more widely in its commencement, duration and amount and hence, it is more important in influencing the nature of water balance. Variations in these two important climatic elements have been analysed year to year by water balance procedures for a 16-year period (1957-72) for 14 selected meteorological stations, namely, Mussoorie, Dehra Dun, Mukteswar, Roorkee, Bahraich, Nautanwa, Agra, Jhansi, Bareilly, Kanpur, Allahabad, Gorakhpur, Varanasi and Ballia, which are situated in different parts of the state.

Figure 8.1 shows clearly a great variation in rainfall of the state from year to year and from place to place, whereas variation in potential evapotranspiration from year to year is less significant, but with respect to latitude and altitude its difference is larger between the U.P. Himalaya and the southern upland.

VARIATION IN RAINFALL

Rainfall dispersion diagram has been drawn to assess more rational variability of rainfall and to classify rainfall regimes (Fig. 8.1). Monthly values are plotted to depict rainfall dispersion for each month during the period (1957-72). These three sets of variables (amount, month and year) are defined simply in terms of the relative positions of the quartiles and median values, which can easily be marked. The lower quartiles of each month at every station is less in extent than the upper quartiles. The upper quartiles mark the rainy months of each year whereas the lower quartiles signify the drought prone months. Maximum variation occurs in the upper quartiles due to exceptionally heavy rainfall by the southwest monsoon (June to October). The median values of each month differ considerably from the average values. The average value in every month at each station is invariably lower than the median. It is, however, occasionally observed in the upper quartiles at Agra, Allahabad and Bahraich in June, at Kanpur in June and September, at Varanasi in June, August and October, at Ballia in September and October and at Gorakhpur in October.

The discontinuities in rainfall at any station may be estimated from major, minor and graded breaks between adjacent months. The interquartile bands for May and June at Nautanwa and for June and July at Varanasi clearly depict major breaks. Similarly minor and graded breaks are conspicuous in the months of July to October at Agra, Jhansi, Kanpur, Bareilly, Bahraich, Allahabad, Gorakhpur, Varanasi and Ballia. At other places like Nautanwa, Mukteswar, Mussoorie and Dehra Dun, minor and graded breaks occur from June to October, because of higher discontinuities of rainfall than at other places during this period.

An attempt has been made to analyse the relative position of quartiles and median values for adjacent months to diagnose major, minor and graded breaks as well as to assess the effect of precipitation on the soil moisture recharge and utilisation. For example, the major break occurs at Nautanwa between May and June, thus, soil moisture recharge at this station occurs in the month of June. If it compensates the cumulative water loss till then, the surplus amount after recharging of soil moisture, yields some water surplus even in this month. The relative position of the interquartile bands have direct effect on variation in actual water loss, soil moisture recharge and soil moisture utilisation. Thus, the computation of water balance for the 16-year period helps in determining the nature of water surplus and water deficit in any individual year. In general, water surplus occurs from July to October, but if water balance is computed for every year, some years show no water surplus at all and in some years large amount of water surplus occurs. The rainfall dispersion diagram shows the extreme limits of rainfall in each month of the 16-year period and relative positions of interquartile bands determine the discontinuities of rainfall on monthly basis rather than annual. It is, however, not suitable for estimating a rational variability of rainfall unless at least a 35-year record of rainfall is available.

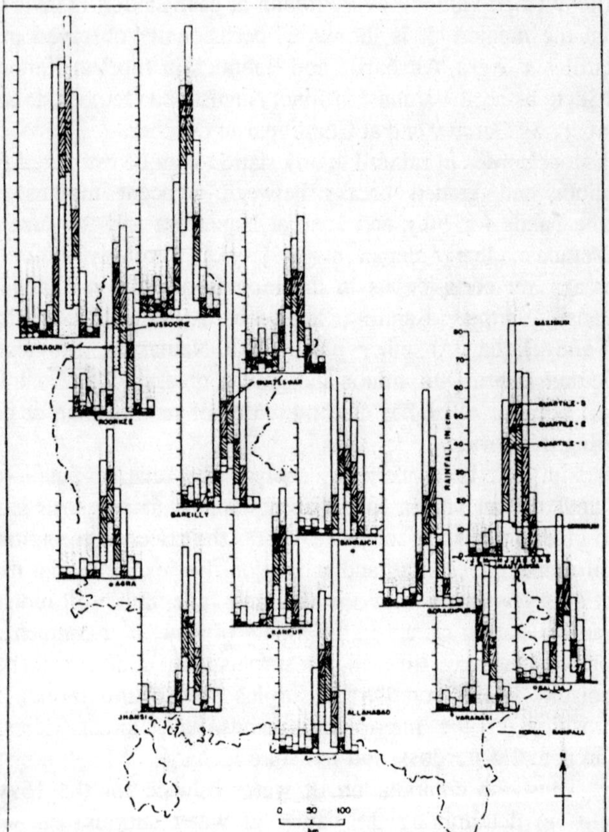

FIG. 8.1: Rainfall Dispersion Diagram, 1957-72

VARIATION IN POTENTIAL EVAPOTRANSPIRATION

Potential evapotranspiration varies slightly from year to year. This variation occurs due to difference in temperature between drought year and rainy year as well as due to intensity and number of cold and heat waves during winter and hot weather seasons respectively. In this study, the annual potential evapotranspiration has been observed to vary from 69.5 to 76.5 cm at Mussoorie, 105.4 to 124.1 at Dehra Dun, 67.8 to 78.0 at Mukteswar, 127.0 to 141.2 at Roorkee, 136.8 to 153.8 at Bahraich, 133.1 to 148.6 at Nautanwa, 143.3 to 157.0 at Agra, 142.7 to 158.2 at Jhansi, 136.0 to 152.7 at Bareilly, 141.0 to 154.1 at Kanpur, 143.2 to 162.9 at Allahabad, 130.8 to 158.2 at Gorakhpur, 136.9 to 156.3 at Varanasi and 138.5 to 156.4 cm at Ballia (Table 8.1). The annual potential evapotranspiration in individual years also varies from place to place following the pattern of its variation in the average annual values.

By comparing precipitation with potential evapotranspiration from year to year on monthly basis, the amount of water surplus and deficit for each year has been determined. Use of the mean monthly data results in a wrong estimation of both water surplus and deficit at all stations within the state, and does not reveal the water problem in its temporal perspective. The reason is, that many of the months have water deficit in some year, surplus in other and storage recharge and utilisation in still other years and hence we must have a complete computation for each individual year. After computing these data for a 16-year period it is found that the amount of both surplus and deficit is larger than that based on average temperature and precipitation data. This computation reveals enough information regarding variations in frequency and intensity of surplus and deficit patterns.

VARIATION IN ACTUAL EVAPOTRANSPIRATION

Actual evapotranspiration generally varies with rainfall. Its actual quantity is equal to that of rainfall if no water surplus occurs. In this study the computed values of actual evapotranspiration during the 16-years period vary from 55.7 to 71.5 cm at Mussoorie, 67.7 to 94.9 at Dehra Dun, 52.4 to 69.9 at Mukteswar, 69.5 to 106.9 at Roorkee, 63.7 to 114.9 at Bahraich, 77.9 to 120.9 at Nautanwa, 48.6 to 97.3 at Agra, 57.2 to 95.5 at Jhansi, 59.3 to 113.5 at Bareilly, 44.4 to 101.2 at Kanpur, 49.1 to 117.0 at Allahabad, 53.5 to 120.0 at Gorakhpur, 60.6 to 109.6 at Varanasi and 59.1 to 106.3 at ballia (Table 8.2).

TABLE 8.1: Annual Potential Evapotranspiration (in cm) 1957-72

Stations	1957	1958	1959	1960	1961	1962	1963	1964	1965	1966	1967	1968	1969	1970	1971	1972
Mussoorie	73.2	76.1	74.5	74.8	73.2	72.3	72.2	74.4	72.2	76.5	71.7	69.6	73.5	72.9	71.8	74.7
Dehra Dun	116.4	124.1	120.8	118.8	117.3	115.5	116.0	115.6	112.5	116.8	113.5	117.1	119.5	110.7	105.4	116.1
Mukteswar	69.2	74.7	78.0	73.6	71.0	72.0	69.9	71.8	70.9	74.8	69.5	70.7	70.1	71.3	67.8	70.1
Roorkee	131.6	141.2	138.0	135.3	130.9	135.4	134.7	130.8	127.4	126.8	127.0	130.8	134.9	132.7	127.9	129.9
Bahraich	136.8	153.7	149.0	149.1	144.7	143.7	148.6	151.0	150.6	152.2	152.3	147.0	153.8	146.4	140.5	147.7
Nautanwa	148.6	146.4	142.3	133.1	145.9	145.4	142.8	141.5	138.1	141.8	135.6	139.9	138.9	139.7	137.8	137.6
Agra	142.6	154.0	148.1	143.4	143.3	143.4	148.6	146.0	150.5	157.0	147.4	146.5	155.6	150.9	144.7	153.4
Jhansi	149.3	158.3	154.3	154.5	150.2	154.0	154.9	150.2	155.0	157.7	149.9	150.4	153.3	148.9	142.7	n.a.
Bareilly	142.3	151.9	147.7	145.3	143.3	144.4	149.6	148.9	149.2	151.5	139.4	136.7	152.7	143.5	136.0	143.8
Kanpur	148.0	155.4	150.9	148.6	145.2	146.1	153.5	154.8	152.7	153.9	148.7	146.0	154.5	148.8	141.0	150.1
Allahabad	153.2	160.3	157.2	153.6	153.6	151.9	154.4	154.9	155.6	162.9	150.6	151.5	159.8	149.3	143.2	152.8
Gorakhpur	149.3	158.2	155.8	157.3	153.1	139.8	149.1	150.1	154.9	160.7	151.8	142.5	152.9	144.0	140.9	146.5
Varanasi	146.8	154.9	148.9	149.3	147.1	147.0	136.9	151.1	150.2	154.1	146.8	146.8	156.3	147.9	144.0	150.0
Ballia	139.4	156.4	156.3	153.6	155.2	150.2	143.3	138.5	146.9	142.1	139.8	143.6	156.4	153.3	148.2	159.9

n.a. = Not availabe

TABLE 8.2: Annual Actual Evapotranspiration (in cm) 1957-72

Stations	1957	1958	1959	1960	1961	1962	1963	1964	1965	1966	1967	1968	1969	1970	1971	1972
Mussoorie	71.5	59.5	57.7	64.8	60.6	65.6	65.2	63.3	56.2	62.6	62.2	55.7	61.4	60.7	68.6	57.4
Dehra Dun	88.3	74.4	94.9	80.0	77.7	69.9	82.9	78.1	77.9	90.6	80.8	84.8	67.7	87.8	88.6	70.9
Mukteswar	58.9	64.4	69.9	63.3	60.6	63.0	67.0	58.8	53.7	61.5	56.1	56.5	60.9	62.8	65.4	52.4
Roorkee	99.7	92.1	105.9	85.5	95.5	99.4	90.8	85.3	79.2	100.3	93.1	69.5	84.0	86.1	100.3	106.9
Bahraich	96.5	104.9	80.5	98.9	102.9	89.9	95.1	100.0	47.4	72.0	63.7	83.9	66.6	114.9	93.6	76.1
Nautanwa	80.1	100.5	111.7	96.3	105.3	105.2	96.4	86.8	78.9	77.9	92.2	100.6	99.1	106.9	120.9	88.2
Agra	88.1	81.0	64.0	75.7	68.5	70.9	70.4	74.8	63.1	74.1	84.9	57.8	74.5	66.0	97.3	48.6
Jhansi	81.1	83.5	85.6	91.0	95.5	68.1	89.8	82.6	74.8	60.5	85.5	57.2	84.0	71.8	74.1	n.a.
Bareilly	83.5	97.1	96.8	105.6	112.1	78.2	83.3	85.4	59.3	96.3	100.6	90.4	88.0	97.7	113.5	90.5
Kanpur	78.5	76.6	80.4	74.0	99.7	67.4	70.2	91.3	64.2	83.7	97.2	74.8	74.1	64.5	101.2	44.4
Allahabad	85.3	72.1	86.8	80.5	93.6	86.3	92.9	85.0	57.2	67.6	94.9	41.9	80.9	105.0	117.0	81.4
Gorakhpur	78.2	84.5	90.7	97.5	103.5	87.2	82.8	86.4	83.2	53.5	69.1	88.3	91.7	102.7	120.0	90.1
Varanasi	85.1	84.7	93.9	81.0	95.1	84.8	85.7	86.2	73.4	68.4	87.3	60.6	100.3	96.4	109.6	65.9
Ballia	71.7	103.3	90.4	90.0	95.7	89.3	87.7	106.3	68.8	63.1	59.1	93.8	86.2	103.5	103.1	66.3

n.a. = Not available

SEASONAL VARIATION OF WATER SURPLUS

During the rainy months (July to October) invariably precipitation exceeds potential evapotranspiration and water surplus occurs. The average water surplus varies from nil to 60 cm in the months of July, August and September. In general the monthly water surplus during the aforesaid period occurs below 5 cm over the plain, 5 to 10 cm over the Tarai, and 15 to 30 cm over the U.P. Himalaya. Over the southern upland, however, a water surplus of 5 to 10 cm occurs in the month of August. Water surplus occurring at Dehra Dun and Mussoorie amounts to 49.8 and 51.6 cm in July and 58.3 and 59.0 cm in August, respectively. Average water surplus is below 5 cm in October over the whole state. Varanasi and Kanpur record no water surplus in October and July respectively (Fig. 8.2A and Table 8.3). Water surplus does occur over the U.P. Himalaya and along the Tarai during winter season but the amount is usually insignificant. The average of 16-year period reveals a surplus of more than 2 cm in January at Mukteswar and Dehra Dun and in February at Mussoorie and Dehra Dun.

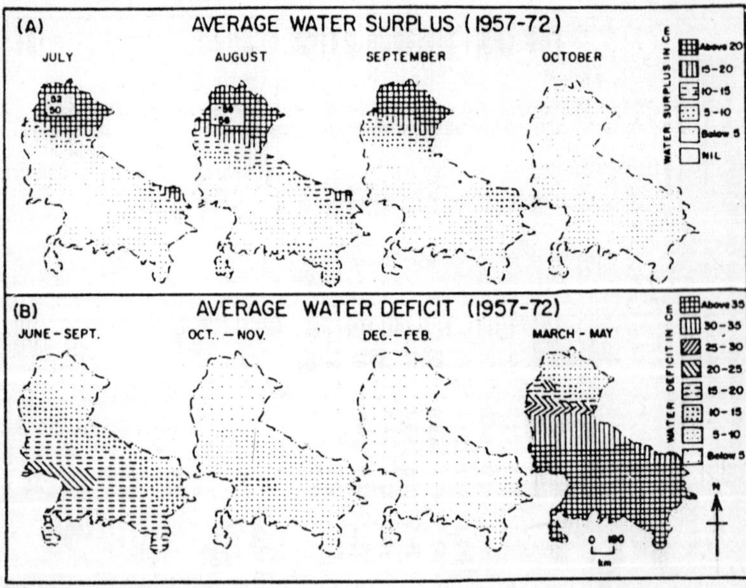

FIG. 8.2: (A) Average Water Surplus, (B) Average Water Deficit

TABLE 8.3: Average Water Surplus and Deficit (1957-72)

Stations	Water Surplus				Water Deficit			
	July	Aug.	Sept.	Oct.	South west mon-soon	Post-mon-soon season	Winter season	Hot weather
Mussoorie	51.6	59.0	22.2	1.3	1.6	2.1	1.0	6.7
Dehra Dun	49.8	58.3	27.0	1.0	6.7	4.5	1.3	22.5
Mukteswar	18.7	21.2	17.5	2.4	1.8	2.4	1.7	5.2
Roorkee	5.6	13.6	5.9	0.2	13.4	3.1	1.0	22.7
Bahraich	0.4	2.8	3.1	1.6	15.6	6.7	3.9	35.1
Nautanwa	20.7	23.3	8.3	2.3	2.5	5.0	3.8	33.0
Agra	0.2	4.4	1.5	0.1	23.9	8.5	3.7	39.4
Jhansi	2.9	6.7	5.0	0.3	16.1	7.4	4.9	40.5
Bareilly	1.7	10.8	4.8	2.8	12.9	5.3	2.8	31.9
Kanpur	0.0	2.4	0.2	0.8	21.1	12.1	4.2	39.0
Allahabad	1.8	6.3	5.9	0.4	18.2	7.4	4.7	39.6
Gorakhpur	7.1	15.6	5.1	0.7	11.1	6.1	5.4	39.9
Varanasi	1.4	4.3	3.3	0.0	16.3	6.6	3.5	37.5
Ballia	1.9	4.1	2.2	0.2	14.1	7.2	4.9	36.8

SEASONAL VARIATION OF WATER DEFICIT

When potential evapotranspiration exceeds precipitation the soil moisture utilisation and water deficit occur. Average water deficit varies from 10 to 25 cm over the plain and the southern uplands whereas it is below 10 cm over the U.P. Himalaya and the Tarai during the southwest monsoon (June to September). The average water deficit is below 10 cm during the post-monsoon period (October-November) and winter period (December to February) except at Kanpur (12.9 cm). It is 35 to 40 cm over most of the plain and the southern upland, 15 to 35 cm over the northwestern plain and the Tarai and 5 to 15 cm over the U.P. Himalaya during hot weather period (March to May) (Fig. 8.2B).

ANNUAL WATER SURPLUS AND DEFICIT PATTERNS IN RAINY YEARS (1961, 1970, 1971)

Annual water surplus and deficit during the 16-year period reveal the rainy years as 1961, 1970 and 1971 with respect to the state as a whole.

In 1961 water surplus varied from 20 to 40 cm over the plain, 40 to 60 cm over the Tarai and about 100 cm over the U.P. Himalaya. It was recorded to be 207.1 and 186.6 cm at Dehra Dun and Mussoorie respectively. Similarly, water deficit varied from 20 to 60 cm over the plain except the southwestern parts where it was 74.8 cm at Agra. It varied from 10 to 40 cm over the U.P. Himalaya due to the effects of elevation except over the high ranges and snowclad areas (Fig. 8.3A). Variation of water surplus in 1970 was 15 to 87.6 cm in the eastern and 15 to 40 cm in the western U.P. with a wide central zone (trending N-S) represented by Bareilly, Agra, Kanpur, and Jhansi, where there was practically no water surplus. It was observed to be about 50 cm over the U.P. Himalaya (120.3 and 138.1 cm at Mussoorie and Dehra Dun along the windward slopes). The annual water deficit during this year was 31.5 to 84.9 cm over the whole of the state except the U.P. Himalaya, where it varied from 8.5 to 22.9 cm (Fig. 8.3B). The year of 1971 was an excessive rainy year, during which the water surplus varied from 100 to 225 cm over the U.P. Himalaya, 40 to 60 cm over the Tarai including the northern parts of the plain and 20 to 40 cm over the southern parts of the state. It was, however, only 4.4, 1.9 and zero at Kanpur, Agra and Bahraich respectively. The amount of water deficit varied from 2.4 to 20.0 cm over the U.P. Himalaya and the Tarai and 20 to 60 cm in the rest of the state except at Jhansi where it was 68.7 cm.

ANNUAL WATER SURPLUS AND DEFICIT IN DROUGHT YEAR (1966, 1968, 1972)

In these years too, water surplus did occur over the U.P. Himalaya and the Tarai along with the northern parts of the plain. It was 40 to 140 cm over the U.P. Himalaya varying with the nature of the terrain and decreased towards the plain. At places like Mussoorie and Dehra Dun the water surplus was observed to be 127.1 and 90.6 cm in 1966, 113.3 and 107.3 cm in 1968 and 120.2 and 137.3 cm in 1972. It was almost nil over most of the plain and the southern upland. Similarly, water deficit increased progressively as we proceed from north to south over the state. Larger water deficit occurred in 1972 at Kanpur (105.7 cm) and Agra (104.8 cm). Water deficit during these drought years varied from 10 to 50 cm over the U.P. Himalaya, 25 to 75 cm in the northern plain including the Tarai and 50 to 100 cm over the rest of the state (Fig. 8.3D, E and F).

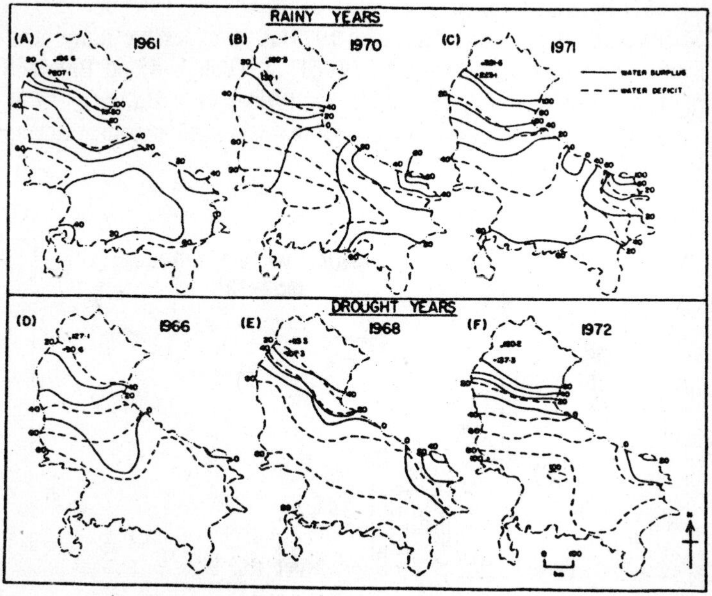

FIG. 8.3: Annual Water Surplus and Deficit: (A) Rainy Years 1961, 1970, 1971, (B) Drought Years 1966, 1968, 1972

WATER SURPLUS AND WATER DEFICIT PATTERNS (1957-72)

Water surplus and water deficit have been computed for 16 years and depicted on maps (Figs. 8.4 and 8.5). These maps show a considerable variation in the patterns of water surplus (Table 8.4) and water deficit (Table 8.5) from year to year in comparison to the average annual values. It is a well known fact that excessive water deficit results in drought and large water surplus exceeding the capacity of the runoff channels causes floods. During many years of this period, some drought was experienced in almost all regions but its intensity varied regionally. Thus special patterns emerging out in excessive rainy and drought years during this 16 year period in different parts of the state have been illustrated by water balance graphs (Figs. 8.6 and 8.7) (Appendices H and I). It may be mentioned here that excessive rainy and drought years are not the same for all stations. The respective years for different stations have been chosen on the merit of the relevant data pertaining to the station concerned after a careful scrutiny of the water balance graphs.

Major parts of the state witnessed 1971 as an excessive rainy year. At Kanpur and Jhansi it occurred in 1961, at Allahabad and Bahraich in 1970, at Agra in 1967, at Bareilly in 1960 and at Ballia in 1964, because rainfall varied more widely from one place to other during

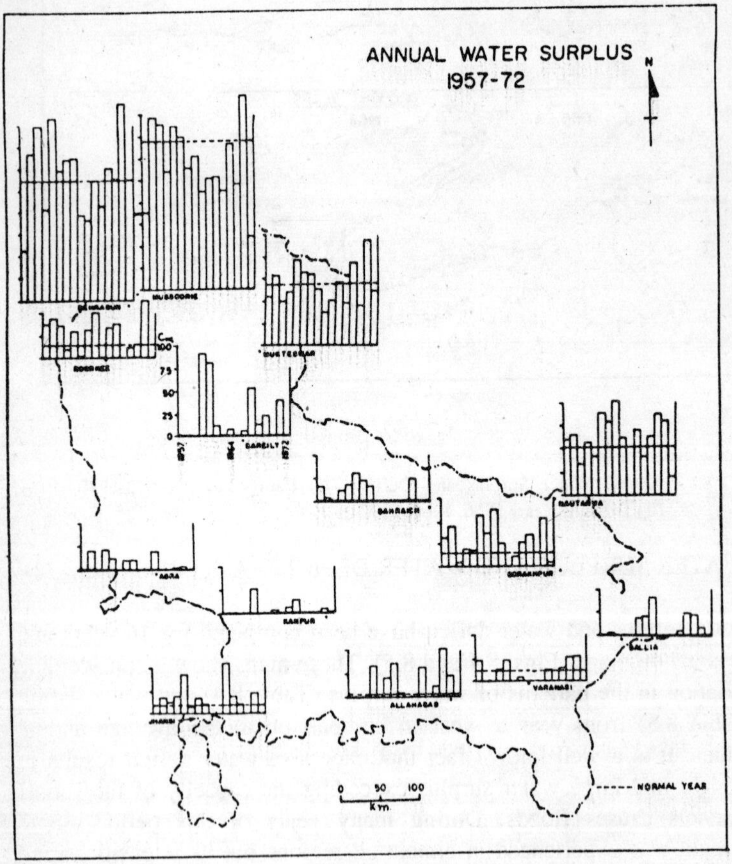

FIG. 8.4: Annual Water Surplus, 1957-72

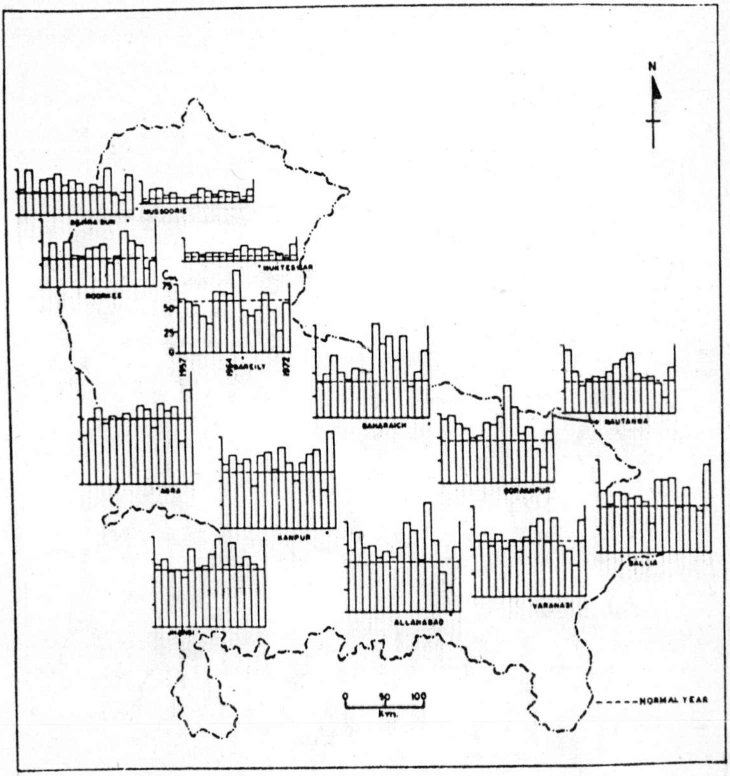

FIG. 8.5: Annual Water Deficit, 1957-72

these typical years. Water surplus occurred during these excessive rainy years in the months of August and September at Ballia, Varanasi, Gorakhpur, Allahabad, Kanpur, Bahraich, Agra, Roorkee and Jhansi over the Ganga plain. It occurred from July to October and from January to March at Dehra Dun, Mussoorie, Mukteswar in the U.P. Himalaya and from July to October at Bareilly and Nautanwa. In general, large water surplus occurred in the month of August. Water recharge took place in the month of June and December in the U.P. Himalaya and during June and July in the Ganga plain. Soil moisture recharge was also witnessed in the winter season in the Tarai region.

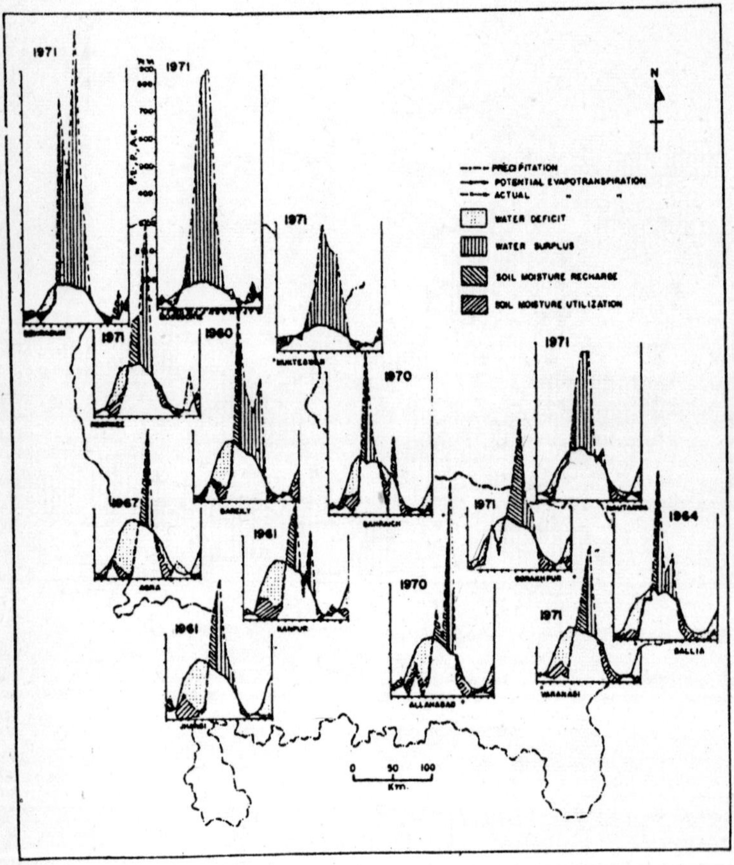

FIG. 8.6: Water Balance for Excessive Rainy Year

Likewise, 1972 was witnessed as excessive drought year, for Mussoorie, Dehra Dun, Agra, Kanpur, Varanasi and Ballia, 1965 for Mukteswar, Bareilly and Bahraich, 1968 for Allahabad and Roorkee and 1966 for Gorakhpur, Jhansi and Nautanwa. In a dry year, rainfall is less than the water need at almost all stations except at Mukteswar, Mussoorie, Dehra Dun, and Nautanwa. I.M.D. has defined drought as

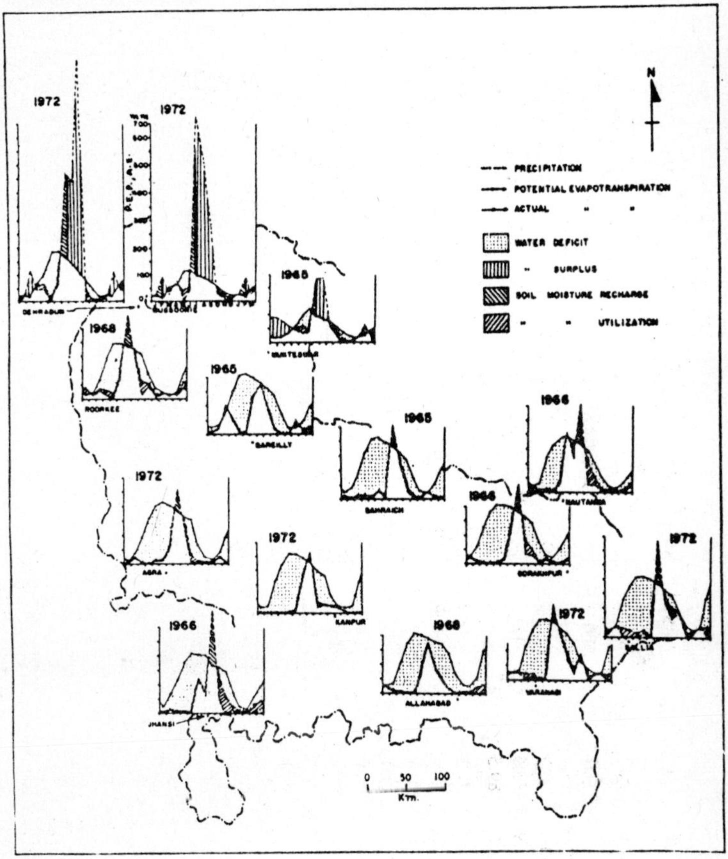

FIG. 8.7: Water Balance for Excessive Drought Year

"a situation occurring in any area when the annual rainfall is less than 75 per cent of the normal". Further, it has defined moderate drought "when rainfall deficit is between 25 per cent and 50 per cent of the normal" and severe drought "when the deficiency is above 50 per cent of the normal".[1] It may be mentioned that "drought is a relative term for, what is drought for one region may be quite normal or even

TABLE 8.4: Annual Water Surplus (in cm), 1957-72

Stations	1957	1958	1959	1960	1961	1962	1963	1964	1965	1966	1967	1968	1969	1970	1971	1972
Mussoorie	86.9	194.5	188.0	134.8	186.6	176.1	138.7	151.5	111.9	127.1	128.4	113.3	166.2	120.3	221.3	120.2
Dehra Dun	56.8	166.4	197.8	103.1	207.1	147.0	160.0	160.7	95.6	90.6	135.7	107.3	150.8	138.1	223.1	137.3
Mukteswar	54.6	78.3	49.8	60.7	96.0	81.6	80.1	63.5	37.2	51.5	71.3	47.0	92.6	43.7	118.6	60.3
Roorkee	38.6	43.7	37.0	10.3	30.5	1.7	31.2	51.7	0.2	34.8	38.0	0.0	12.0	16.1	63.6	0.8
Bahraich	15.9	4.2	0.0	11.7	17.9	30.5	21.9	16.2	0.0	0.0	0.0	0.0	0.0	24.6	0.0	0.0
Nautanwa	59.5	65.2	36.8	56.7	42.5	89.5	79.0	102.3	62.4	0.3	70.8	41.2	59.1	87.8	81.3	19.9
Agra	0.0	21.6	0.0	21.8	12.4	0.0	9.9	9.9	0.0	0.0	19.5	0.0	0.0	1.7	1.9	0.0
Jhansi	9.0	16.7	11.5	17.7	43.2	5.6	16.6	9.8	0.0	0.0	4.5	18.2	10.5	0.0	22.5	n.a.
Bareilly	0.0	0.0	0.0	91.5	81.1	11.4	0.0	7.1	0.0	5.2	53.7	12.4	22.6	0.0	38.6	0.0
Kanpur	0.0	0.0	0.0	0.0	26.2	0.0	0.0	3.3	0.0	6.6	14.8	0.0	0.0	0.0	4.4	0.0
Allahabad	0.0	0.0	0.0	34.4	13.4	21.3	36.3	4.3	0.0	0.0	32.5	0.0	11.6	52.3	23.2	0.0
Gorakhpur	31.2	44.6	4.6	19.1	18.5	55.5	35.9	61.1	25.6	0.0	10.9	25.5	34.2	33.6	52.0	3.7
Varanasi	22.3	0.0	0.0	12.8	23.1	0.0	4.3	0.0	0.0	0.0	12.8	0.0	14.5	25.8	26.5	0.0
Ballia	0.0	3.1	0.0	0.0	0.0	18.0	20.4	43.9	0.0	0.0	0.0	5.2	23.7	18.5	8.3	0.0

n.a. = Not available

TABLE 8.5: Annual Water Deficit (in cm), 1957-72

Stations	1957	1958	1959	1960	1961	1962	1963	1964	1965	1966	1967	1968	1969	1970	1971	1972
Mussoorie	1.7	16.6	196.8	10.0	12.6	6.7	7.0	10.7	16.0	13.9	9.5	13.9	12.1	12.2	3.2	17.3
Dehra Dun	28.1	49.7	25.9	38.8	39.5	45.6	33.1	37.5	34.6	26.2	32.7	32.3	51.8	22.9	16.8	45.2
Mukteswar	10.3	10.3	8.1	10.3	10.4	9.0	2.9	13.0	17.2	13.3	13.4	14.2	9.2	8.5	2.4	17.7
Roorkee	31.9	49.1	32.1	49.8	35.4	36.0	43.9	45.5	48.2	25.5	33.9	61.3	50.9	46.6	21.0	29.6
Bahraich	40.3	48.8	68.5	50.2	41.8	53.8	63.5	51.0	103.2	80.2	88.6	63.1	87.2	31.5	46.9	71.6
Nautanwa	68.5	45.9	30.6	36.8	40.6	40.2	46.2	54.7	59.2	63.9	43.4	39.3	39.8	32.8	16.9	49.4
Agra	54.1	73.0	85.1	57.5	74.8	72.5	78.2	71.2	87.4	82.9	62.5	88.7	81.1	84.9	47.4	104.8
Jhansi	68.2	74.8	68.7	63.5	54.7	85.9	65.1	67.6	80.2	97.2	64.1	93.2	69.3	77.1	68.6	n.a.
Bareilly	58.8	54.8	50.9	39.7	31.2	66.2	66.2	63.5	89.9	55.2	38.8	46.3	64.7	45.8	22.5	53.0
Kanpur	69.5	78.8	70.5	74.6	45.5	78.7	83.3	63.5	88.5	70.2	51.5	71.2	80.4	84.3	39.8	105.7
Allahabad	67.9	88.2	70.4	73.1	60.0	65.6	61.5	69.9	98.4	95.3	55.7	109.6	78.9	44.3	26.2	71.4
Gorakhpur	71.1	73.7	65.1	59.8	49.6	52.6	66.3	63.7	71.7	107.2	82.7	54.2	61.2	41.3	20.9	56.4
Varanasi	61.7	70.2	55.0	68.3	52.0	62.2	51.2	64.9	76.8	85.5	59.5	86.2	56.0	51.5	34.4	84.1
Ballia	67.7	53.1	65.9	63.6	59.5	60.9	55.6	32.4	78.1	79.0	80.7	49.8	70.2	49.8	45.1	93.6

n.a. = Not available

favourable to another which, therefore, leads to the assessment of drought and its effect not in terms of rainfall but only through water balance criteria.[2] Therefore, it should not be assumed that the more humid regions are free from drought. Drought is of greater intensity in the normal dry region but even a humid region suffers from severe moisture deficit in some years. A comparative study of water deficit in a normal year, in a rainy year and in a dry year reveals some interesting fact. In an excessive rainy year, water deficit occurred at all stations from 47 to 69 per cent of the normal except in the southwest semi-arid part like Agra, Kanpur, Jhansi and Allahabad where it varied between 77 and 80 per cent. Water deficit occurring in an excessive year was 337, 275, 261, 228 and 204 per cent of the normal at Mukteswar, Mussoorie, Bahraich, Gorakhpur and Allahabad respectively. It varied between 147 and 200 per cent at the remaining stations. Thus, it is more important to study the severity of drought during excessive dry years.

VARIATION IN MOISTURE INDEX

Moisture index is derived from water surplus and water deficit [(I_m = *100 (S-D/PE)]*. Due to the changing of annual water surplus and deficit, variation in moisture index for each station is found to occur from year to year. When rainfall exactly equals the potential evapotranspiration in every month meaning thereby that water is available just as needed, there is neither water deficit nor water excess and the climate in neither moist nor dry, but it can only be a hypothetical case. As water deficit becomes larger with respect to potential evapotranspiration the climate becomes more arid, on the other hand, if water surplus becomes larger the climate becomes more humid. Water surplus and water deficit have been computed for each individual year and on their basis, the moisture index has been determined. The former affects it positively and the latter negatively. Yearly moisture index (I_m) has been plotted against the corresponding years (Fig. 8.8). It exhibits appreciable deviation from the normal (Table 8.6).

The perhumid and humid climate groups belong to the humid category. Moisture index of Mukteswar, Mussoorie and Dehra Dun vary between humid and perhumid and hence this region has the least probability of facing any water deficit problem. Of the other stations, Nautanwa represents Moist subhumid, Agra, Jhansi, Kanpur and Ballia Semiarid and the remaining stations Dry subhumid category. The

TABLE 8.6: Variation in Moisture Index in 1957-72

Stations	1957	1958	1959	1960	1961	1962	1963	1964	1965	1966	1967	1968	1969	1970	1971	1972
Mussoorie	116.4	233.8	229.8	166.8	237.7	234.3	182.4	190.3	132.8	148.0	165.8	142.8	249.7	148.3	303.8	137.8
Dehra Dun	24.7	94.0	142.3	54.1	142.8	87.8	109.4	106.6	54.2	55.1	90.8	64.1	82.8	104.1	195.7	79.3
Mukteswar	64.0	91.0	53.5	64.5	120.6	100.8	110.4	70.3	28.2	51.1	83.3	46.4	119.0	49.4	171.4	60.8
Roorkee	5.1	-3.8	3.6	-29.2	-4.4	-25.3	-9.4	4.7	-37.7	6.5	3.2	-46.9	-28.8	-23.0	33.3	-22.2
Bahraich	-17.8	-29.0	-46.0	-25.8	-16.5	-16.2	-21.7	-23.0	-68.5	-52.8	-58.2	-42.9	-56.7	-4.7	-33.4	-48.5
Nautanwa	-6.1	13.2	4.4	15.0	-1.4	33.9	22.8	33.6	2.3	-44.9	20.2	1.4	13.9	39.4	46.7	-21.4
Agra	-37.9	-33.4	-57.5	-32.1	-43.6	-50.6	-46.0	-42.0	-58.1	-52.8	-27.2	-60.6	-52.1	-55.1	-31.5	-68.3
Jhansi	-39.7	-45.7	-37.1	-29.6	-6.6	-52.1	-31.3	-38.5	-51.7	-61.6	-6.4	-49.9	-38.4	-51.8	-32.3	n.a.
Bareilly	-41.3	-36.1	-34.5	35.7	34.8	-38.0	-44.3	-37.9	-60.3	-33.0	10.7	-24.8	-27.6	-31.9	11.8	-36.9
Kanpur	-47.0	-50.7	-46.7	-50.2	-13.3	-53.9	-54.3	-38.9	-58.0	-41.3	-24.7	-48.8	-50.0	-56.7	-25.1	-70.4
Allahabad	-44.3	-55.0	-44.8	-25.2	-30.3	-29.2	-16.3	-42.4	-63.2	-58.5	-15.4	-72.2	-42.1	5.4	-2.1	-46.7
Gorakhpur	-26.7	-18.4	-38.8	=25.9	-20.3	-2.3	-20.3	-1.7	-29.8	-66.7	-47.3	-20.2	=17.7	-5.4	22.1	-36.0
Varanasi	-26.8	-45.3	-36.9	-37.2	-19.7	-42.3	-34.3	-42.7	-51.1	-55.5	-31.8	-58.9	-26.6	-17.4	-5.5	-56.1
Ballia	-48.6	-32.0	-42.2	-41.4	=38.7	-28.6	-9.6	8.3	-53.2	-55.6	-57.7	-31.1	-29.4	-20.4	-24.8	-58.4

n.a. = Not available

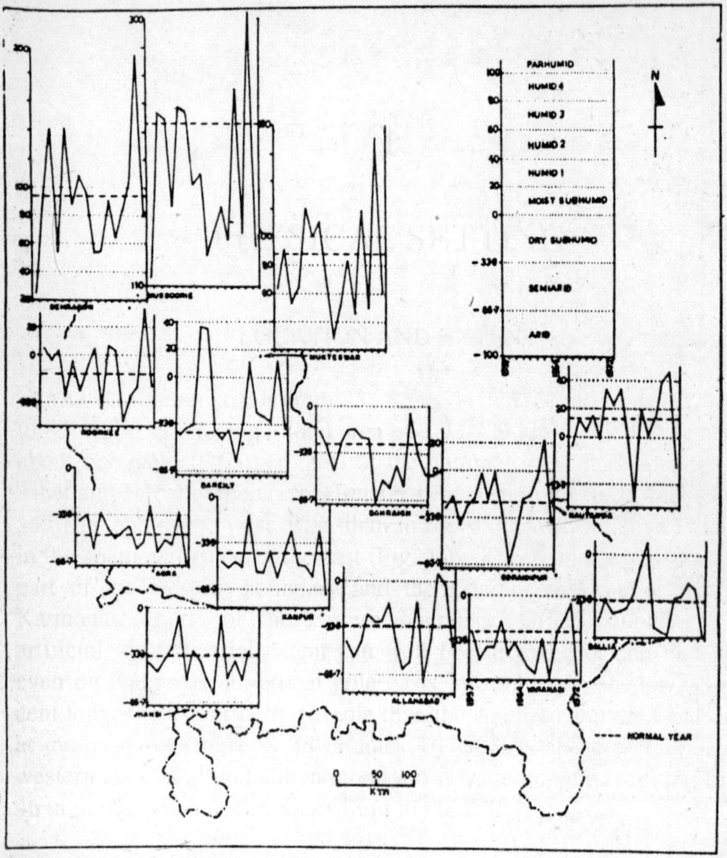

FIG. 8.8: Variation in Moisture Index, 1957-72

computed values of moisture index shift to the humid side in a rainy year and to the arid side in a dry year (Fig. 8.8). Fluctuations of moisture index in comparison to normal values are found greater in the U.P. Himalaya, the Tarai and northern parts of the Ganga plain, e.g., at Roorkee, Bareilly, Bahraich and Gorakhpur where rainfall is more than that in the south and central parts of the state.

NOTES

1. Singh, T.: *Drought Prone Areas in India: Aspect of Identification and Development Strategy*, New Delhi (1978), p. 27.
2. Subrahmanyam, V.P. et al.: *Delimitation of Drought Prone Areas — Techniques and Methodology*, All India Symposium on Drought Prone Areas of India, Tirupati (1978), p. 1.

9

SUMMARY AND CONCLUSION

UTTAR PRADESH is a unique state in India, comprising parts of all the three major physical divisions of the Indian Republic. It varies in elevation from 46 m at its eastern tip to 7,618 m in the Nanda Devi peak. With respect to geological structure the state again covers parts of three major structural divisions of India, comprising part of the Himalaya folds, the Ganga trough and the stable Deccan plateau. On the basis of its soil types and vegetation, the moisture holding capacity varies from 50 mm in the U.P. Himalaya to 400 mm in the southeastern upland.

The spatio-temporal variations in distribution of climatic elements, i.e., temperature, pressure and wind, relative humidity, rainfall and frequency of atmospheric disturbances (hail, thunder, fog, duststorm and squall) have thoroughly been studied and described. January and July temperature distribution over the state have been shown with the help of isotherms, which give an erroneous picture because they eliminate the effect of altitude. However, annual and diurnal ranges of temperature are found to be generally lower at mountainous stations than those over the plain this error might have cropped because of the higher values of temperature anomaly at the mountainous stations (Appendix B). Similarly, they are lower in eastern part than in the western part because of increase in the degree of continentality. Mean annual temperature anomalies are greater in the U.P. Himalaya than over the Ganga plain and the southern upland. Similarly the anomalies in the southern upland are comparatively greater than those in the Ganga plain. General distribution of high pressure in January and low pressure in July are affected by western disturbances in the winter months and Bay depressions in the monsoon months. Consequently, they control the wind direction and its speed. The pattern of relative humidity exhibits two maxima one in winter and the other in the rainy season interspersed by two minima in spring and autumn. It is higher in morning than in the evening. Along some slopes in the U.P. Himalaya, however it is higher in the evening except in the dry summer months of May and June. The annual distribution of rainfall varies from windward side to the leeward side of the ranges and from the hill stations to the intermountain valleys. It also varies with elevation except in the northern snowclad ranges of the U.P. Himalaya. Rainfall is about 70 cm at Agra in the Ganga plain from where it increases towards the Tarai, the eastern parts of the plain and southern upland. The variability of rainfall is however inversely related to the amount of rainfall. The annual variability of rainfall varies from 15 to 35 per cent. As the state receives 70 to 90 per cent of its annual rainfall during the southwest monsoon season, the variability of rainfall during this season indicates the trend of its annual variability. It varies from 20 to 40 per cent. The pattern of rainy days are similar to that of annual rainfall and its number varies from 30 to 100. The state is also affected by hailstorms, thunderstorms, fog, duststorm and squall (Table 2.2) during different seasons.

Climographs vividly reveal the fact that comfortable weather condition prevails in the months of March, April, October and November over the Ganga plain, whereas the months of May and June provide quite comfortable living condition over the ranges of the U.P. Himalaya. Hythergraphs are helpful in assessing suitability of the climatic condition for different crops. It must, however, be remembered that by providing irrigational facilities man has changed the nature's dictates to a considerable extent.

The I.M.D. has recognized four seasons, i.e., southwest monsoon, post-monsoon, winter and hot weather. The standard deviation in onset dates of southwest monsoon is about 6 days. The normal dates of onset are 15th June in east U.P. and 25th June in west U.P. Sometimes, it advances about a week later in the U.P. Himalaya than in the west plain. It withdraws from the state between the third week of September and the first week of October. The oscillation of seasonal trough plays a major role in distribution and intensity of southwest monsoon rain, which extends from Punjab to West Bengal. The state receives nearly 70 to 90 per cent of its annual rainfall during this season. Average amount of rainfall varies from 120 to 200 cm in the U.P. Himalaya, 80 to 120 in the east U.P. and 50 to 80 cm in the west U.P. It is above 200 cm at Nainital, Mussoorie and Munsiari in the U.P. Himalaya. The rainfall intensity and number of rainy days vary from 15 to 30 mm and 30 to 70 days respectively. High percentage of relative humidity occurs during this season, which varies from 60 to 90 per cent.

During post-monsoon season, the rainfall is about 5 per cent. The amount of rainfall varies from 5 to 10 cm at Nainital and east U.P. and below 5 cm over the rest of the state. The rainfall intensity varies from 15 to 25 mm per rainy day and mean number of rainy days is about three. Mean relative humidity varies from 45 to 75 per cent during this season.

During winter season, cloud and rainfall occur due to the western disturbances, the intensity of which gradually decreases form west to east. The percentage of winter rainfall is 2 to 10 of the annual total over the state. The amount of rainfall is 10 to 20 cm in the U.P. Himalaya, 5 to 10 cm in the northwestern part of the Ganga plain and below 5 cm in the rest of the state. The rainfall intensity varies from 10 to 20 mm per day and mean rainy days from 2 to 15 days. The mean relative humidity varies from 50 to 75 per cent and decreases towards the southwest.

During hot weather season, the percentage of rainfall is up to ten. The amount of rainfall varies from 5 to 20 cm, which occurs due to western disturbances. The rainfall intensity is 5 to 15 mm per day and mean number of rainy days varies form 2 to 10 days. The mean relative humidity ranges from 23 to 48 per cent. During this season, storms frequently occur during the hot day time and mitigate the air temperature without any rainfall. Strong and hot westerly winds, locally known as *"Loo"*, are also characteristic features of summer afternoons.

Cold waves are generally experienced form November to March, but in some years they have also been experienced in April and October. On the other hand, heat waves are experienced from April to July. Nearly 54 per cent of their incidence is recorded in June, when east U.P. experiences the maximum number of severe heat waves.

Climatic classification has been recognized on the basis of Lang's Rain factor, De Mortonne's Aridity index, Koppen's scheme, Kira's Index of Moist and Dry climate and Thornthwaite's scheme. Among these, Koppen's and Thornthwaite's schemes have gained popularity in geographical literature. According to Koppen's scheme, the wet and dry tropical climate (Aw) emerges over the southern upland. Low latitude steppe climate (Bsh) occurs over Aligarh, Agra and Mainpuri districts. Humid mesothermal (Cw) climate emerges over the rest of the state. With the help of potential evapotranspiration, water surplus and water deficit of 39 stations, four relevant information of water balance have been obtained for Thornthwaite's Rational classification of climate. These are, Moisture index, Seasonal variation of effective moisture, Thermal efficiency and Summer concentration of Thermal efficiency. On their basis the state has been divided into 17 subclimatic regions. The superiority of Thornthwaite's approach over those of other climatologists can be pleaded on the basis of the fact that he took into consideration two important elements, namely, potential evapotranspiration and moisture index which play a significant role in the energy exchange between the earth surface and the atmosphere and are of considerable importance in determining the types of climate.

Microclimatological variations in a hypothetical urban and in mountainous areas have been dealt. As a case study, a comparison of climatic data of Varanasi through its urban fringe with those of its surrounding rural areas clearly reveals the progressive contrast between the two extremes. Industrialization in the urban areas polluted the air and consequently affects the resultant isolation and radiation to some extent. It also contributes extra nuclei for development of water droplets around them. The dense urban construction makes the system a better heat store, the surface more waterproof and the air to stagnate longer. Thus, temperature, relative humidity, frequency of fog and rainfall are comparatively higher in the centre of the city than in the surrounding suburbs and rural environments. Similarly, temperature in the U.P. Himalaya varies from 3.2 (Landaur 2289 m) to 14.6°C (Kalsi 609 m) in the coldest month of January and from 18.9 (Munsiari 2300 m) to 30.0°C (Kalsi 609 m) in the hottest months of May and June, showing the effect of altitude. It also differs according to the gradient and direction of slopes. South and southwest facing slopes receive maximum effective insolation than the north and northeast facing slopes because the sun remains south of this area throughout the year. Both diurnal and annual ranges of temperature decrease on ascending from the plain up to an elevation of 2400 m, beyond which they again increase, their greatest values being attained at the highest elevations probably due to more rarefied air and paucity of impurities. Like temperature, relative humidity also differs considerably on the windward and leeward slopes as well as at different elevations. On the windward slopes of the Himalayas, high percentage of relative humidity has been observed due to adiabatic cooling of the ascending wind and consequent condensation of moisture. On the leeward slopes temperature rises and humidity decreases due to katabatic warming. It is also greater over the forestclad slopes and valleys in comparison to the steep and bare mountain slopes and valleys. The maximum amount of rainfall occurs along the middle zone of the windward slopes. At elevations much above the saturation level, the decrease in absolute humidity reduces the amount of rainfall. Similarly, there is a sharp contrast in the amount of rainfall along the windward and the leeward slopes of the Himalayan ranges. The southwest monsoon rain also gradually diminishes in amount as we move along the Himalayan range from east to west as well towards the ranges situated north of the outer range.

The concept of potential evapotranspiration, its estimation and spatial distribution have been studied according to the Thornthwaite's and Penman's formulae. It is defined by Thornthwaite, as the optimum water loss from surface completely covered with vegetation and supplied with abundant water at all times. Its quantity depends only on the amount of solar energy received by the surface and its resulting temperature. He also gave a comprehensive statement that AE = PE as long as there is any available water and AE = O at soil moisture below the wilting point or at zero degree air temperature. Variation in altitude, latitude and distance from sea as well as seasonal characteristics of rainfall affect to a considerable extent the distribution of potential evapotranspiration. The isopleth of 142.5 cm runs almost east-west in the featureless middle parts of the Ganga plain, but in the U.P. Himalaya, the high elevation and diverse topography cause the computed values of P.E. to vary from 69.5 to 122.5 cm. A comparative study of the results based on Thornthwaite's and Penman's formulae reveal the fact that the computed values of annual P.E. are almost the same in both the cases, but Penman's monthly values of P.E. are higher in winter and lower in summer than those of Thornthwaite.

Climatic water balance and its different components, i.e., actual evapotranspiration, soil moisture utilization, soil moisture recharge, water deficit and water surplus have been dealt, on the basis of which the state has revealed single as well as multiple wet and dry seasons. In the U.P. Himalaya, the periods of soil moisture utilization as well as that of water deficit, to some extent, occur during March to May and October-November. The soil moisture recharge and water surplus periods occur from June to September and December to February. Monthwise computation of water balance for Ballia, Pipri and western Semiarid parts do not show any water surplus although the water need is less than precipitation during the monsoon months of June to September. It happens so because, the surplus rain is totally spent in recharging the soil moisture. In the Moist subhumid Tarai region, soil moisture recharge occurs in January and February in winter and again starts in June during the summer. Water surplus occurs during the period July to October and water deficit from March to June and in November. In the Dry subhumid eastern region, soil moisture recharge occurs in June and July and water surplus occurs only in the wet and warm months of August and September.

The U.P. Himalaya has considerable water surplus, i.e., 33.9 to 199.3 cm. The Tarai along with the submontane zone has marked water surplus of 25 to 50 cm. In the central and eastern parts, the annual water surplus amounts to 25 cm. In contrast, a region with a considerable water deficit (60 to 80 cm) lies in the western part. There are also large water deficit (40 to 60 cm) areas, extending into the central and eastern, 'Dry subhumid' parts. The U.P. Himalaya and the Tarai have an annual water deficit of 0.3 to 34.7 cm where water need is comparatively less. The distribution of Humidity and Aridity indices is similar to the water surplus and deficit patterns. Moisture index varies from above 100 to below –33.3, i.e., Perhumid to Semiarid. The Moisture Adequacy index values exceeding 80 per cent are found in the U.P. Himalaya and less than 40 per cent at Agra and its surroundings. A comparative study of water balance on the basis of Thornthwaite and Penman reveals the fact that Penman's P.E. curve declines assuming a concave shape during the southwest monsoon season, whereas, the Thornthwaite's P.E. curve during the same period assumes a convex shape. Consequently, they cause conspicuous differences in the amounts of water surplus and water deficit. A notable feature of water balance in this study area obtained by Penman's method is that the water deficit during the southwest monsoon months is lower but in winter months it is higher in comparison to the water balance obtained by applying Thornthwaite's method. Thus, Penman's results are quite close to the reality as it is experienced in these seasons.

Yearwise water balance for a 16-year period (1957-72) of 14 selected stations have been computed, which exhibit a great variation in rainfall from year to year causing considerable variations in water surplus and deficit in the state during that period. In the average values of these computations, it is found that both surplus and deficit are larger than those based on average temperature and precipitation data, because many of the months have deficit in some year, surplus in other and storage rechange and utilization in still other years. Considerable annual variation in the pattern of water surplus and deficit have been observed in comparison to the normal annual values. The special patterns emerging out in excessive Rainy and Drought years during this 16-year period in different parts of the state have been illustrated by water balance graphs. It must be mentioned here that excessive Rainy and Drought years are not the same for all stations. The respective years for different stations have been chosen on the merit of the relevant data pertaining to the station concerned after a careful scrutiny of the water balance graphs.

A comparative study of water deficit in a normal year, in an excessive Rainy year and in an excessive Dry year reveals some interesting facts. In an excessive Rainy year water deficit at all stations was 47 to 80 per cent of the normal. Water deficit occurring in an excessive Dry year was 337, 275, 261, 228 and 204 per cent of the normal at Mukteswar, Mussoorie, Bahraich, Gorakhpur and Allahabad respectively. It was found to vary from 147 to 200 per cent at the remaining stations.

Due to the changing pattern of annual water surplus and deficit, variation in moisture index for each station is also observed variation in moisture index in comparison to the respective normal values is found to be greater in the U.P. Himalaya, the Tarai and northern parts of the Ganga plain, where rainfall is more than that in the south and central parts of the state.

CONCLUSION AND SUGGESTIONS FOR FUTURE PROSPECTS

The climatic characteristics as well as water surplus and deficit patterns have been studied analytically in the preceding chapters. These studies have exposed a number of problems that the state faces today. The main problems have been diagnosed and relevant remedial measures have been suggested below to solve these problems as far as practicable.

(i) In the mountainous area there is paucity of meteorological and even rain recording stations, particularly at higher elevations. At the same time there is no arrangement to measure the depth of snow pack. It is necessary to establish many more meteorological stations at the crest line and along the slopes to expose the micro variation of climatic elements in this area. It is also advisable to measure the areas and average depth of snow-packs to assess the potential melt water.

(ii) The main rivers causing serious floods from west to east are the Yamuna, the Ganga, the Ramganga, the Ghaghra and the Gandak. The windward slopes of the U.P. Himalaya receive huge amount of orographic rain. Sharp breaks in the longitudinal profile of these rivers cause floods. As a result of this sharp break of slope in the piedmont region, the rivers, transporting down huge load of sediments, start aggrading their valleys, thus, reducing their water carrying capacity. The result is that the surplus rain water spreads over the banks.

Afforestation along the mountain slopes and construction of artificial levees on both sides of rivers have been a very old practice of arresting the overflow of water within the channel. But these provide only temporary relief. The construction of multipurpose dams have been quite effective during recent years in impounding the excess water and regulating the water flow. In the long run even the future of such dams seems to be uncertain because of:

(i) Silting problem of the reservoirs.

(ii) Cracking of dams due to the great strain during the rainy season. Intermittent dredging and construction of inundation canals are two effective measures for controlling floods. By dredging (a) the reservoirs can be deepended to hold more water and the channels rendered capable of discharging the extra flood water, (b) the course may be straightened for increasing the gradient and consequent easy flow, (c) the dredged material may be used for artificial embankment on both the banks. Inundation canals can intake some of the flood water and help in checking floods. They may also provide irrigation water to distant fields on the fertile plain.

(iii) In the mountainous and foothill regions the water yield is larger, because of higher precipitation and low potential evapotranspiration. The variation from year to year is also large. On the other hand, the Semiarid plains usually have lower precipitation, higher potential evapotranspiration and larger deficit. The management of stream flow improvement and afforestation in the catchment areas are needed to minimize silt transport, runoff and soil erosion and will promote better moisture storage. As a result of this operation the vegetal cover will also develop closer to the possible optimum.

(iv) Amount of water available for groundwater recharge has locational advantage in its use. It is potential for domestic, smaller urban and industrial uses as well as for local supplementary irrigation. The cost of pumping may be high but it can be controlled if users take advantage of telluric water movements near the surface and select aquifers that have adequate recharge potential.

(v) Specific crops are grown in different parts and in different seasons. They reflect the moisture pattern and are affected by variation in rainfall amount from year to year. An improvement in crop's productivity is possible by soil moisture conservation in dry years and flood control in rainy years. Irrigation and crop scheduling in a general way are possible if monthly data of potential evapotranspiration and precipitation are taken into account. The aspect of water supply through lift canals can be managed on the basis of a long term observation of stream flow.

(vi) Regarding other uses of fresh water, i.e., for urban, industrial, hydroelectric power, navigational, commercial, fishing and recreational uses, adequate information about water balance pattern is needed, because proper water management requires such data. The imbalance of regional distribution is accentuated by greater demand for water in the plain area. Construction of lift canals from major perennial streams is necessary to supplement the existing modes of water supply.

Improved and proper water management for the state is essential. Its objectives should include an improvement in stream flow yield, regime and quality, reduction of soil erosion, control of flood damages and proper provision for direct economic uses of water in the area. Some of these merits are inter-competitive and hence all cannot be achieved simultaneously. Therefore, a proper water management policy involves an assessment of comparative merit and priority to be given to more desired and beneficial requirements like generation of hydro-electricity and irrigation even at the cost of others like fishing, recreation etc.

APPENDICES

APPENDIX A

Temperature in °C

Stations		Jan.	Feb.	Mar.	Apr.	May	Jun.	Jul.	Aug.	Sept.	Oct.	Nov.	Dec.	Annual	No. of Years
Mussoorie	c.	6.4	7.8	11.7	16.5	20.0	20.3	18.4	17.9	17.1	14.9	11.6	8.5	14.2	30
Dehra Dun	c.	12.6	14.8	19.4	24.6	28.9	29.5	26.8	26.1	25.5	22.2	17.5	14.0	21.8	30
Mukteswar	c.	5.8	7.3	11.1	15.4	18.7	19.0	17.6	17.6	16.5	14.2	11.2	8.2	13.6	30
Nainital	c.	6.0	8.6	12.3	15.8	19.,7	20.5	19.2	18.8	17.9	14.2	10.7	8.3	14.3	10
Joshimath	c.	6.8	8.4	12.3	17.0	20.0	22.2	21.1	20.8	19.5	17.0	11.3	8.8	15.4	10
Tehri	c.	12.0	14.5	19.2	24.4	28.3	29.8	28.2	28.1	27.0	20.3	17.3	12.9	21.8	7
Mukhim	c.	7.8	8.8	12.7	17.4	20.5	21.4	20.4	20.3	19.4	16.4	12.6	9.9	15.6	7
Munsiari	c.	6.5	8.3	8.4	13.8	15.5	18.9	18.2	17.4	16.5	13.6	10.5	8.5	13.1	3
Roorkee	c.	13.4	15.8	20.9	26.7	31.5	32.2	29.4	28.7	27.9	24.1	18.3	14.4	23.1	30
Najibabad	c.	13.7	16.1	21.6	27.0	31.1	32.3	29.8	28.5	28.1	24.1	18.8	15.0	24.2	10
Pantnagar	c.	13.1	15.8	20.6	25.7	29.1	30.6	28.6	28.5	27.6	23.7	18.3	14.2	23.0	10
Bahraich	c.	15.7	18.3	23.7	29.2	32.7	32.3	29.7	29.2	28.7	26.4	21.0	16.9	25.3	30
Nautanwa	c.	15.8	18.1	23.4	28.4	30.8	30.8	29.1	28.6	28.4	25.8	20.9	16.8	24.7	10
Meerut	c.	14.2	17.1	22.5	28.1	32.4	33.4	30.3	29.2	28.8	25.2	19.6	15.6	24.8	10
Aligarh	c.	14.7	17.4	22.8	28.9	33.8	34.2	30.8	29.4	29.0	21.2	20.6	16.0	25.3	30
Agra	c.	14.8	18.0	23.8	29.7	34.5	35.0	30.4	29.3	28.9	26.2	20.6	16.2	25.7	30
Agra (Obsy.)	c.	14.3	17.3	23.2	28.8	32.6	34.8	30.5	28.9	28.0	26.4	22.4	15.4	25.2	10
Mainpuri	c.	15.2	18.1	23.7	29.7	34.2	34.8	30.9	29.5	29.2	26.5	20.9	16.5	25.8	30
Jhansi	c.	16.7	19.8	25.5	31.1	35.7	34.8	29.7	28.3	28.3	26.4	21.4	17.4	26.3	30
Orai	c.	15.7	19.1	25.1	30.4	34.9	34.5	29.8	28.3	28.6	26.4	20.8	16.9	25.9	10

Contd.

Contd. APPENDIX-A

Stations		*Jan.*	*Feb.*	*Mar.*	*Apr.*	*May*	*Jun.*	*Jul.*	*Aug.*	*Sept.*	*Oct.*	*Nov.*	*Dec.*	*Annual*	*No. of Years*
Banda	c.	16.7	19.9	25.7	31.2	35.5	35.1	30.2	28.9	29.0	26.6	21.2	17.4	26.4	10
Bereilly	c.	15.3	17.9	23.3	29.1	33.2	33.3	30.0	29.1	28.9	25.9	20.5	16.6	25.2	30
Kheri	c.	15.6	18.6	23.8	29.3	32.8	32.3	29.3	29.0	28.8	25.9	20.8	17.0	25.3	10
Hardoi	c.	15.3	18.4	23.9	29.3	33.2	33.6	30.1	29.3	29.1	16.1	20.6	16.8	25.5	10
Lucknow	c.	16.1	18.9	24.6	30.1	33.8	33.7	30.1	29.3	29.1	26.3	20.5	17.0	25.9	30
Lucknow (Obsy.)	c.	15.1	18.0	23.7	29.4	33.4	33.2	30.0	29.1	28.9	25.8	19.8	16.0	25.2	10
Kanpur (Obsy.)	c.	15.7	18.5	24.5	30.2	34.5	34.3	30.2	29.0	28.8	26.2	20.6	16.4	25.2	30
Kanpur	c.	14.5	17.8	24.2	29.5	34.1	33.8	30.1	29.0	28.7	25.9	20.2	15.7	25.3	10
Fatehpur	c.	16.2	18.9	24.7	30.4	34.8	34.5	30.3	29.0	28.9	26.4	20.9	16.8	26.0	30
Allahabad	c.	16.4	19.2	25.2	30.7	34.8	34.4	30.1	29.1	29.0	26.5	21.1	17.1	26.1	30
Gonda	c.	15.6	18.1	23.7	29.3	32.8	32.2	29.6	29.1	28.7	26.1	20.6	16.6	25.2	30
Gorakhpur	c.	16.5	19.1	24.9	29.9	32.5	31.6	29.6	29.3	29.1	26.9	21.6	17.6	25.7	30
Azamgarh	c.	16.5	19.1	24.9	29.9	32.5	31.6	29.6	29.3	29.1	26.9	21.6	17.6	25.9	10
Varanasi	c.	16.5	19.3	25.3	30.5	34.3	33.7	30.0	29.1	29.1	26.6	21.0	17.1	26.0	30
Varanasi (Obsy.)	c.	16.2	19.4	20.1	30.1	34.2	33.6	29.7	29.2	28.9	26.7	20.8	17.1	26.0	10
Faizabad	c.	15.6	18.8	23.9	28.9	32.9	31.9	29.7	28.9	28.3	25.4	20.6	16.4	25.0	10
Sultanpur	c.	15.6	19.1	24.7	30.1	32.6	32.6	29.5	29.0	29.0	26.3	21.2	16.5	25.5	10
Ballia	c.	16.9	18.6	23.9	27.8	33.6	33.3	30.1	29.2	29.2	26.6	20.8	17.5	25.6	10
Pipri	c.	19.40	23.0	27.8	32.3	35.8	34.2	29.8	29.0	29.5	29.4	24.7	20.5	28.4	10

a. Mean maximum b. Mean Minimum c. Mean temperature

Source: Calculated on the basis of the maximum and minimum temperature (°C) given in the "climatological Tables of observation of India 1931-1960 I.M.D. pp. 85-118 and 349-364 (1967) published by Government of India"

APPEDIX B

Temperature in °C

Stations	*Terrestrial Coordinates*		*Alti-tude*	*(Reduced to sea level)*			*Ann-ual range*	*Diur-nal range*	*Temp-era-ture ano-maly*
	N	*E*	*m*	*Jan.*	*Jul.*	*Ann-ual*			
Mussoorie	30°27′	78°05′	2266	19.9	31.9	27.7	13.9	7.6	7.4
Dehra Dun	30°19′	78°02′	683	16.7	30.9	25.9	16.9	12.0	5.6
Mukteswar	29°28′	79°39′	2311	19.7	31.5	27.5	12.3	8.1	6.7
Nainital	29°23′	79°27′	2024	18.1	31.3	26.4	14.5	8.1	5.6
Joshimath	30°33′	79°35′	1861	18.0	32.3	26.6	15.4	9.8	6.4
Tehri	30°22′	78°29′	770	16.6	32.8	26.4	17.8	14.0	6.1
Mukhim	30°35′	78°28′	1800	18.6	31.2	26.4	13.6	8.8	6.2
Munsiari	30°0′	80°17′	2300	20.3	32.0	26.9	12.4	8.7	6.4
Roorkee	29°51′	77°53′	274	15.0	31.0	25.2	18.8	13.2	4.6
Najibabad	29°37′	78°21′	267	15.3	31.4	25.8	18.6	12.6	5.1
Pantnagar	29°0′	79°30′	250	14.6	30.1	24.5	17.5	14.8	3.5
Bahraich	27°34′	81°36′	124	16.4	30.4	26.0	17.0	12.4	4.0
Nautanwa	27°26′	83°26′	99	16.4	29.7	25.3	15.0	13.0	3.4
Meerut	29°01′	77°43′	222	15.5	31.6	26.1	19.6	12.9	5.1
Aligarh	27°53′	78°04′	188	15.8	31.9	26.4	19.5	13.2	4.4
Agra (Obsy.)	27°10′	78°02′	167	15.8	30.3	26.7	20.2	13.3	4.3
Agra	27°12′	78°0′	157	15.7	31.8	26.6	19.6	13.2	4.2
Mainpuri	27°14′	79°03′	157	16.1	30.4	26.7	19.6	14.1	4.3
Jhansi	25°27′	78°35′	251	18.2	31.2	27.8	19.0	13.1	4.5
Orai	25°59′	79°30′	139	16.5	30.6	26.7	19.2	13.5	3.7
Banda	25°28′	80°22′	121	17.4	30.9	27.1	18.9	13.0	3.8
Bareilly	28°22′	79°24′	173	16.3	31.0	26.2	18.0	12.6	4.5
Kheri	27°54′	80°48′	147	16.5	30.2	26.2	18.5	12.5	4.2
Hardoi	27°23′	80°10′	142	16.1	30.9	26.3	18.2	12.9	4.4
Lucknow (Obsy.)	26°45′	80°53′	113	16.8	30.8	26.6	17.7	12.9	3.9
Lucknow	26°52′	80°56′	120	16.8	30.2	26.4	17.2	12.8	3.7
Kanpur (Obsy.)	26°26′	80°22′	126	16.5	31.0	26.0	18.8	11.8	3.2
Kanpur	26°28′	80°21′	140	16.6	31.0	25.7	18.7	12.9	3.7
Fatehpur	25°56′	80°50′	124	16.9	31.0	26.0	18.6	12.9	3.6
Allahabad	25°27′	81°44′	98	17.0	30.7	26.1	18.4	12.6	3.4
Gonda	27°08′	81°58′	110	16.3	30.3	25.2	17.2	12.7	3.4
Gorakhpur	26°45′	83°22′	77	17.0	30.2	25.7	16.0	11.4	3.5
Azamgarh	26°03′	83°13′	78	17.0	30.1	25.9	17.2	12.7	3.4
Varanasi (Obsy.)	25°27′	82°52′	85	17.0	20.5	26.0	17.8	12.4	3.2
Varanasi	25°18′	83°01′	76	16.9	30.4	26.0	17.8	13.4	3.1
Faizabad	26°47′	82°08′	102	16.2	30.3	25.0	16.2	13.5	2.9
Sultanpur	26°16′	82°05′	97	16.2	30.1	25.5	17.0	13.3	3.2
Ballia	25°44′	84°10′	64	17.3	30.5	25.6	16.7	13.3	2.8
Pipri	24°0′	82°45′	300	21.1	31.6	28.0	16.4	12.3	5.8

Source: Terrestrial coordinate altitude and temperature data were taken from the "climatological Tables of observation of India 1931-1960 I.M.D. pp. 85-118 and 349-364 (1967) Published by Government of India" reduced to sea level, Diurnal and Annual ranges of temperature and temperature anomaly have been calculated by the Authoress.

APPENDIX C

Seasonal and Annual Distribution of Average Relative Humidity (%)

Stations	*Monsoon season*	*Post-monsoon season*	*Winter season*	*Hot-weather season*	*Annual*
Mossoorie	86.75	64.50	65.67	47.83	68.00
Dehra Dun	73.13	65.00	65.33	38.50	61.00
Mukteswar	83.88	56.76	53.00	42.33	61.00
Nainital	79.38	64.25	61.50	42.83	62.00
Joshimath	76.50	56.00	48.00	40.00	55.00
Tehri	65.75	63.75	66.61	41.83	60.00
Mukhim	79.50	61.75	53.83	43.50	60.00
Munsiari	78.75	59.25	46.50	33.40	54.00
Roorkee	68.60	61.00	65.17	35.67	58.00
Najibabad	69.00	57.50	69.67	36.67	58.00
Pantnagar	73.50	65.00	67.67	45.67	64.00
Bahraich	76.63	63.25	65.83	39.17	60.00
Nautanwa	76.75	68.00	72.00	46.33	67.00
Meerut	73.00	66.00	72.00	44.33	64.00
Aligarh	64.88	49.75	56.67	31.50	52.00
Agra (Obsy.)	64.75	48.25	53.67	28.67	50.60
Agra	64.75	48.25	53.67	28.67	50.50
Mainpuri	68.25	56.00	61.67	33.17	56.00
Jhansi	67.00	45.00	48.17	23.17	48.00
Orai	70.13	51.00	55.67	34.83	54.50
Banda	72.63	59.50	61.17	32.83	58.00
Bareilly	69.70	60.00	64.17	34.00	58.00
Kheri	76.63	70.25	68.83	42.17	65.00
Hardoi	72.63	66.25	66.33	37.17	61.00
Lucknow (Obsy.)	73.00	64.50	64.83	35.33	60.00
Lucknow	72.00	64.00	64.33	39.83	58.00
Kanpur (Obsy.)	69.50	57.50	60.33	31.00	54.00
Kanpur	69.50	57.50	60.33	30.50	55.50
Fatehpur	71.50	58.75	59.83	33.33	57.00
Allahabad	69.75	58.00	60.50	27.66	55.00
Gonda	75.38	68.00	66.00	38.83	63.00
Gorakhpur	75.50	64.75	64.83	40.33	62.00
Azamgarh	78.75	70.25	71.33	43.33	67.00
Varanasi (Obsy.)	72.63	60.00	61.00	29.50	57.00
Varanasi	72.50	60.50	60.50	32.50	57.00
Faizabad	75.75	71.00	68.33	39.67	64.00
Sultanpur	73.75	60.56	59.67	33.33	58.00
Ballia	75.13	63.75	62.17	36.67	59.00
Pipri	78.75	75.50	73.00	43.66	70.00

Source: Average annual Relative Humidity, seasonal distribution of Relative Humidity have been calculated by the Authoress on the basis of the monthly data available in the "climatological Tables of observation of India 1931-1960 I.M.D. pp. 85-118 and 349-364 (1967) published by Government of India".

APPENDIX D

Climatic Indices

Stations	*Lang's Rainfactor*	*De Mortonne's Index of Aridity*	*Kira's Index of moist and dry climate*	*Thornthwaite T.E. (cm)*	*S.C.T.E. (cm)*
Mossoorie	166.8	97.9	18.86	72.7	41.95
Dehra Dun	106.1	72.7	13.53	119.2	45.04
Mukteswar	99.9	57.6	11.22	72.3	40.53
Nainital	181.7	106.9	20.62	72.6	41.59
Joshimath	87.2	52.9	10.13	78.6	42.11
Tehri	47.9	32.8	6.10	122.5	45.79
Mukhim	109.5	66.7	12.77	78.3	41.63
Munsiari	204.9	116.2	22.81	69.5	42.02
Roorkee	49.3	34.7	6.41	134.0	44.55
Najibabad	51.8	36.6	6.85	134.9	44.11
Pantnagar	60.9	42.2	7.87	125.3	44.37
Bahraich	46.5	33.3	6.13	147.0	40.54
Nautanwa	65.6	46.7	8.61	142.3	40.27
Meerut	33.7	24.1	4.45	140.3	43.41
Aligarh	30.7	22.0	4.10	147.2	42.46
Agra (Obsy.)	29.8	21.4	3.90	147.9	41.85
Agra	26.5	19.04	3.51	147.8	41.95
Mainpuri	29.5	21.3	3.92	150.4	40.89
Jhansi	38.0	27.5	5.06	152.7	39.68
Orai	40.3	29.1	5.34	153.6	39.48
Banda	38.0	27.6	5.05	155.3	39.28
Bareilly	42.4	30.3	5.58	144.5	41.11
Kheri	44.7	32.1	5.91	147.1	41.06
Hardoi	45.5	32.7	5.99	151.1	40.04
Lucknow (Obsy.)	38.3	27.6	5.10	149.4	40.63
Lucknow	39.7	28.6	5.26	148.9	40.09
Kanpur (Obsy.)	34.9	25.0	4.52	148.3	41.04
Kanpur	43.9	31.6	5.81	148.4	41.04
Fatehpur	34.04	24.6	4.52	151.8	40.51
Allahabad	39.4	28.4	5.22	152.6	39.97
Gonda	51.4	36.5	6.77	144.7	40.98
Gorakhpur	49.6	35.7	6.56	150.5	38.94
Azamgarh	44.5	29.2	5.35	152.1	39.38
Varanasi (Obsy.)	42.8	29.0	5.67	150.9	39.36
Varanasi	38.3	27.7	5.08	151.9	39.76
Faizabad	46.2	33.0	6.09	143.8	40.26
Sultanpur	43.2	31.0	5.71	148.9	40.03
Ballia	38.0	46.7	5.03	147.2	40.83
Pipri	37.8	27.8	5.09	170.5	35.37

T.E. = Thermal efficiency; S.C.T.E. = Summer concentration of Thermal efficiency

Source: Calculated by the authoress (The calculations made by the authoress are on the basis of the modus operendi explained in the book).

APPENDIX E

Climatic Water Balance Tables (Thornthwaite)

Stations	Jan.	Feb.	Mar.	Apr.	May	Jun.	Jul.	Aug.	Sept.	Oct.	Nov.	Dec.	Annual
Dehra Dun (Moisture Holding Capacity 5.0 cm)													
T°C	12.6	14.8	19.4	24.6	28.9	29.5	26.8	26.1	25.5	22.2	17.5	14.0	
P.E.	1.9	2.6	6.5	12.0	18.4	18.6	16.6	14.7	12.7	8.8	4.3	2.1	119.2
P.	6.2	5.9	4.0	1.7	4.2	18.6	72.1	76.5	32.7	7.1	0.5	1.9	231.4
P.-P.E.	4.3	3.3	-2.5	-10.3	-4.2	0.0	55.5	61.8	20.0	-1.7	-3.8	-0.2	112.2
St.	5.0	5.0	3.0	0.3	0.1	0.1	5.0	5.0	5.0	3.5	1.6	1.5	
Δ St.	3.5	0.0	-2.0	-2.7	-0.2	0.0	4.9	0.0	0.0	-1.5	-1.9	-0.1	
A.E.	1.9	2.6	6.0	4.4	4.4	18.6	16.6	14.7	12.7	8.6	2.4	2.0	94.9
W.D.	0.0	0.0	0.5	7.6	14.0	0.0	0.0	0.0	0.0	0.2	1.9	0.1	24.3
W.S.	0.8	3.3	0.0	0.0	0.0	0.0	50.6	61.8	20.0	0.0	0.0	0.0	136.5
Mussoorie (Moisture Holding Capacity 5.0 cm)													
T°C	6.4	7.8	11.7	16.5	20.0	20.3	18.4	17.9	17.1	14.9	11.6	8.5	
P.E.	1.6	1.8	4.3	7.1	10.6	10.5	9.4	8.6	7.1	5.6	3.5	2.6	72.7
P.	6.6	6.7	6.4	3.0	4.5	18.9	72.7	75.5	32.3	6.5	0.8	3.1	237.0
P.-P.E.	5.0	4.9	2.1	-4.1	-6.1	8.4	63.3	66.9	25.2	0.9	-2.7	0.5	164.3
St.	5.0	5.0	5.0	2.1	0.6	5.0	5.0	5.0	5.0	5.0	2.8	3.3	
Δ St.	1.7	0.0	0.0	-2.9	-1.5	4.4	0.0	0.0	0.0	0.0	-2.2	-0.5	
A.E.	1.6	1.8	4.3	5.9	6.0	10.5	9.4	8.6	7.1	5.6	3.0	2.6	66.4
W.D.	0.0	0.0	0.0	1.2	4.6	0.0	0.0	0.0	0.0	0.0	0.5	0.0	6.3
W.S.	3.3	4.9	2.1	0.0	0.0	4.0	63.3	66.9	25.2	0.9	0.0	0.0	170.6

Stations	Jan.	Feb.	Mar.	Apr.	May	Jun.	Jul.	Aug.	Sept.	Oct.	Nov.	Dec.	Annual
Nainital (Moisture Holding Capacity 5.0 cm)													
T°C	6.0	8.6	12.3	15.8	19.7	20.5	19.2	18.8	17.9	14.2	10.7	8.3	
P.E.	1.4	2.1	4.6	6.8	10.2	10.8	9.2	9.2	7.7	5.3	3.2	2.1	72.6
P.	11.8	4.4	5.4	2.5	7.6	27.3	76.9	56.1	33.1	30.5	0.7	3.5	259.8
P.-P.E.	10.4	2.3	0.8	-4.3	-2.6	16.5	67.7	46.9	25.4	25.2	-2.5	1.4	187.2
St.	5.0	5.0	5.0	2.0	1.1	5.0	5.0	5.0	5.0	5.0	3.0	4.4	
Δ St.	0.6	0.0	0.0	-3.0	-0.9	3.9	0.0	0.0	0.0	0.0	-2.0	1.4	
A.E.	1.4	2.1	4.6	5.5	3.5	10.8	9.2	9.2	7.7	5.3	2.7	2.1	69.1
W.D.	0.0	0.0	0.0	1.3	1.7	0.0	0.0	0.0	0.0	0.0	0.5	0.0	3.5
W.S.	9.8	2.3	0.8	0.0	0.0	12.6	67.7	46.9	25.4	25.2	0.0	0.0	190.7
Nautanwa (Moisture Holding Capacity 15.0 cm)													
T°C	15.8	18.1	23.4	28.4	30.8	30.8	29.1	28.6	28.4	25.8	20.9	16.8	
P.E.	2.2	3.4	9.6	16.1	19.5	19.4	18.4	17.3	15.4	12.4	5.9	2.7	142.3
P.	2.0	1.3	1.8	2.8	2.9	27.5	44.3	46.0	20.7	12.0	0.4	0.5	162.2
P.-P.E.	-0.2	-2.1	-7.8	-13.3	-16.6	8.1	25.9	28.7	5.3	-0.4	-5.5	-2.5	19.9
St.	8.4	7.2	4.3	1.7	0.6	8.7	15.0	15.0	15.0	14.6	10.0	8.4	
Δ St.	0.0	-1.2	-2.9	-2.6	-1.1	8.1	5.3	0.0	0.0	-0.4	-4.6	-1.6	
A.E.	2.0	2.5	4.7	5.4	4.0	19.4	18.4	17.3	15.4	12.4	5.0	2.1	108.6
W.D.	0.2	0.9	4.9	10.7	15.5	0.0	0.0	0.0	0.0	0.0	0.9	0.6	33.7
W.S.	0.0	0.0	0.0	0.0	0.0	0.0	19.6	28.7	5.3	0.0	0.0	0.0	53.6

Stations	Jan.	Feb.	Mar	Apr.	May	Jun.	Jul.	Aug.	Sept.	Oct.	Nov.	Dec.	Annual
Pantnagar (Moisture Holding Capacity 15.0 cm)													
T°C	13.1	15.8	20.6	25.7	29.1	30.6	28.6	28.5	27.6	23.7	18.3	14.2	
P.E.	1.0	2.1	7.2	13.2	18.3	19.1	18.2	17.3	14.8	9.1	3.7	1.8	125.3
P.	4.3	2.2	1.7	0.9	4.3	14.5	40.9	33.9	27.2	8.5	0.5	1.2	140.1
P.-P.E.	3.3	0.1	-5.5	12.3	-14.0	-4.6	22.7	16.6	12.4	-0.6	-3.2	-0.1	14.8
St.	14.7	14.8	10.2	4.4	1.7	1.4	15.0	15.0	15.0	14.4	11.5	11.4	
Δ St.	3.3	0.1	-4.6	-5.8	-2.7	-0.3	13.6	0.0	0.0	-0.6	-2.9	-0.1	
A.E.	1.0	2.1	6.3	6.7	7.0	14.8	18.2	17.3	14.8	9.1	3.4	1.3	102.0
W.D.	0.0	0.0	0.9	6.5	11.3	4.3	0.0	0.0	0.0	0.0	0.3	0.0	23.3
W.S.	0.0	0.0	0.0	0.0	0.0	0.0	9.1	16.6	12.4	0.0	0.0	0.0	38.1
Najibabad (Moisture Holding Capacity 15.0 cm)													
T°C	13.7	16.1	21.6	27.0	31.1	32.3	29.8	28.5	28.1	24.1	18.8	15.0	
P.E.	1.6	2.6	7.3	14.9	19.8	20.3	19.4	17.4	15.5	9.7	4.3	2.1	134.9
P.	5.3	2.6	2.4	1.2	1.7	10.4	40.4	31.4	18.0	10.2	0.5	1.2	125.3
P.-P.E.	3.7	0.00	-4.9	-13.7	-18.1	-9.9	21.0	14.0	2.5	0.5	-3.8	-0.9	-9.6
St.	14.6	14.6	10.5	4.1	1.2	0.6	15.0	15.0	15.0	15.0	11.5	10.9	
Δ St.	3.7	0.0	-4.1	-6.4	-2.9	-0.6	14.4	0.0	0.0	0.0	-3.5	-0.6	
A.E.	1.6	2.6	6.5	7.6	4.6	11.0	19.4	17.4	15.5	9.7	4.0	1.8	101.7
W.D.	0.0	0.0	0.8	7.3	15.2	9.3	0.0	0.0	0.0	0.0	0.3	0.3	33.2
W.S.	0.0	0.0	0.0	0.0	0.0	0.0	6.6	14.0	2.5	0.5	0.0	0.0	23.6

Stations	Jan.	Feb.	Mar.	Apr.	May	Jun.	Jul.	Aug.	Sept.	Oct.	Nov.	Dec.	Annual
Tehri (Moisture Holding Capacity 5.0 cm)													
T°C	12.0	14.5	19.2	24.4	28.3	29.8	28.2	28.1	27.0	20.3	17.3	12.9	
P.E.	1.6	2.4	6.2	12.0	17.7	19.0	19.4	17.1	14.2	6.8	4.3	1.8	122.5
P.	6.4	4.7	4.5	2.0	3.5	5.1	30.2	31.0	21.8	9.0	0.7	1.7	120.6
P.-P.E.	4.8	2.3	-1.7	-10.0	-14.2	-13.9	10.8	13.9	7.6	2.2	-3.6	-0.1	-1.9
St.	5.0	5.0	3.5	0.4	0.0	0.0	5.0	5.0	5.0	5.0	2.3	2.3	
Δ St.	2.7	0.0	-1.5	-3.1	-0.4	0.0	5.0	0.0	0.0	0.0	-2.7	0.0	
A.E.	1.6	2.4	6.0	5.1	3.9	5.1	19.4	17.1	14.2	6.8	3.4	1.7	86.7
W.D.	0.0	0.0	0.2	6.9	13.8	13.9	0.0	0.0	0.0	0.0	0.9	0.1	35.8
W.S.	2.1	2.3	0.0	0.0	0.0	0.0	5.8	13.9	7.6	2.2	0.0	0.0	33.9
Mukteswar (Moisture Holding Capacity 5.0 cm)													
T°C	5.8	7.3	11.1	15.4	18.7	19.0	17.6	17.3	16.5	14.2	11.2	8.2	
P.E.	1.6	2.1	4.3	7.1	9.9	10.1	9.3	8.5	7.4	5.9	3.7	2.4	72.3
P.	5.8	5.6	4.5	3.4	5.1	14.3	33.2	33.8	19.7	7.9	0.4	2.3	136.0
P.-P.E.	4.2	3.5	0.2	-3.7	-4.8	4.2	23.9	25.3	12.3	2.0	-3.3	-0.1	63.7
St.	5.0	5.0	5.0	2.3	0.8	5.0	5.0	5.0	5.0	5.0	2.5	2.4	
Δ St.	2.6	0.0	0.0	-2.7	-1.5	4.2	0.0	0.0	0.0	0.0	-2.5	-0.1	
A.E.	1.6	2.1	4.3	6.1	6.6	10.1	9.3	8.5	7.4	5.9	2.9	2.4	67.2
W.D.	0.0	0.0	0.0	1.0	3.3	0.0	0.0	0.0	0.0	0.0	0.8	0.0	5.1
W.S.	1.6	3.5	0.2	0.0	0.0	0.0	23.9	25.3	12.3	2.0	0.0	0.0	68.8

Stations	Jan.	Feb.	Mar.	Apr.	May	Jun.	Jul.	Aug.	Sept.	Oct.	Nov.	Dec.	Annual
Joshimath (Moisture Holding Capacity 5.0 cm)													
T°C	6.8	8.4	12.3	17.0	20.0	22.2	21.1	20.8	19.5	17.0	11.3	8.8	
P.E.	1.4	1.8	4.0	7.1	9.9	11.6	11.2	10.3	8.3	8.2	2.9	1.9	78.6
P.	7.3	9.5	14.1	6.0	2.2	12.1	30.1	28.3	14.9	3.5	2.7	3.6	134.3
P.-P.E.	5.9	7.7	10.1	-1.1	-7.7	0.5	18.9	18.0	6.6	-4.7	-0.2	1.7	55.7
St.	5.0	5.0	5.0	4.0	0.8	1.3	5.0	5.0	5.0	1.8	1.8	3.5	
Δ St.	1.5	0.0	0.0	-1.0	-3.2	0.5	3.7	0.0	0.0	-3.2	0.0	1.7	
A.E.	1.4	1.8	4.0	7.0	5.4	11.6	11.2	10.3	8.3	6.7	2.7	1.9	72.3
W.D.	0.0	0.0	0.0	0.1	4.5	0.0	0.0	0.0	0.0	1.5	0.2	0.0	6.3
W.S.	4.4	7.7	10.1	0.0	0.0	0.0	15.2	18.0	6.6	0.0	0.0	0.0	62.0
Pipri (Moisture Holding Capacity 40.0 cm)													
T°C	19.4	23.0	27.8	32.8	35.8	34.2	29.8	29.0	29.5	29.4	24.7	20.5	
P.E.	4.5	7.7	15.1	18.4	21.0	20.5	18.8	17.5	16.2	15.7	9.6	5.5	170.5
P.	1.9	1.5	1.2	0.8	1.3	13.9	28.9	30.9	19.4	4.6	0.6	0.8	105.8
P.-P.E.	-2.6	-6.2	-13.9	-17.6	-19.7	-6.6	10.1	13.4	3.2	-11.1	-9.0	-4.7	-64.7
St.	14.9	12.8	9.0	5.8	3.5	3.0	13.1	26.5	29.7	22.5	17.9	16.0	
Δ St.	-1.1	-2.1	-3.8	-3.2	-2.3	-0.5	10.1	13.4	3.2	-7.2	-4.6	-1.9	
A.E.	3.0	3.6	5.0	4.0	3.6	14.4	18.8	17.5	16.2	11.8	5.2	2.7	105.8
W.D.	1.5	4.1	10.1	14.4	17.4	6.1	0.0	0.0	0.0	3.9	4.4	2.8	64.7
W.S.	0.0	0.0	0.0	0.0	0.0	0.0	0.0	0.0	0.0	0.0	0.0	0.0	0.0

Stations	*Jan.*	*Feb.*	*Mar.*	*Apr.*	*May*	*Jun.*	*Jul.*	*Aug.*	*Sept.*	*Oct.*	*Nov.*	*Dec.*	*Annual*
Jhansi (Moisture Holding Capacity 20.0 cm)													
T°C	16.7	19.8	25.5	31.1	35.7	34.8	29.7	28.3	28.3	26.4	21.4	17.4	
P.E.	2.8	4.5	12.1	17.9	21.1	20.9	18.6	16.8	15.3	13.1	6.3	3.3	152.7
P.	1.7	1.4	1.1	0.3	0.8	10.5	29.8	30.9	19.0	3.2	0.5	0.7	99.9
P.-P.E.	-1.1	-3.1	-11.0	-17.6	-20.3	-10.4	11.2	14.1	3.7	-9.9	-5.8	-2.6	-51.8
St.	7.5	6.4	3.7	1.5	0.5	0.4	11.6	20.0	20.0	12.1	9.0	7.9	
Δ St.	-0.4	-1.1	-2.7	-2.2	-1.0	-0.1	11.2	8.4	0.0	-7.9	-3.1	-1.1	
A.E.	2.1	2.5	3.8	2.5	1.8	10.6	18.6	16.8	15.3	11.1	3.6	1.8	90.5
W.D.	0.7	2.0	8.3	15.4	19.3	10.3	0.0	0.0	0.0	2.0	2.7	1.5	62.2
W.S.	0.0	0.0	0.0	0.0	0.0	0.0	0.0	5.7	3.7	0.0	0.0	0.0	9.4
Kanpur (Moisture Holding Capacity 25.0 cm)													
T°C	15.7	18.5	24.5	30.2	34.5	34.3	30.2	29.0	28.8	26.2	20.6	16.4	
P.E.	2.2	3.7	10.5	17.3	21.2	20.7	19.0	17.5	15.6	12.8	5.4	2.4	148.3
P.	2.3	1.6	0.9	0.5	0.6	6.8	20.8	28.6	20.2	4.3	0.8	0.8	88.2
P.-P.E.	0.1	-2.1	-9.6	-16.8	-20.6	-13.9	1.8	11.1	4.6	-8.5	-4.6	-1.6	-60.1
St.	10.1	9.2	6.2	3.2	1.4	0.8	2.6	13.7	18.3	12.8	10.7	10.0	
Δ St.	0.1	-0.9	-3.0	-3.0	-1.8	-0.6	1.8	11.1	4.6	-5.5	-2.1	-0.7	
A.E.	2.2	2.5	3.9	3.5	2.4	7.4	19.0	17.5	15.6	9.8	2.9	1.5	88.2
W.D.	0.0	1.2	6.6	13.8	18.8	13.3	0.0	0.0	0.0	3.0	2.5	0.9	60.1
W.S.	0.0	0.0	0.0	0.0	0.0	0.0	0.0	0.0	0.0	0.0	0.0	0.0	0.0

Stations	Jan.	Feb.	Mar.	Apr.	May	Jun.	Jul.	Aug.	Sept.	Oct.	Nov.	Dec.	Annual
Agra (Moisture Holding Capacity 20.0 cm)													
T°C	14.8	18.0	23.8	29.7	34.5	35.0	30.4	29.3	28.9	26.2	20.6	16.2	
P.E.	1.7	3.4	9.6	17.0	21.3	21.1	19.5	17.6	16.0	12.7	5.4	2.4	147.9
P.	1.6	0.9	1.1	0.5	1.0	6.0	21.0	26.3	15.2	2.4	0.2	0.4	76.6
P.-P.E.	-0.1	-2.5	-8.5	-16.5	-20.3	-15.1	1.5	8.7	-0.8	-10.3	-5.2	-2.0	-71.3
St.	4.0	3.5	2.3	1.0	0.4	0.2	1.7	10.4	9.9	5.9	4.5	4.0	
Δ St.	0.0	-0.5	-1.2	-1.3	-0.6	-0.2	1.5	8.7	-0.5	-4.0	-1.4	-0.5	
A.E.	1.6	1.4	2.3	1.8	1.6	6.2	19.5	17.6	15.7	6.4	1.6	0.9	76.6
W.D.	0.1	2.0	7.3	15.2	19.7	14.9	0.0	0.0	0.3	6.3	4.0	1.5	71.3
W.S.	0.0	0.0	0.0	0.0	0.0	0.0	0.0	0.0	0.0	0.0	0.0	0.0	0.0
Meerut (Moisture Holding Capacity 25.0 cm)													
T°C	14.2	17.1	22.5	28.1	32.4	33.4	30.3	29.2	28.8	25.2	19.6	15.6	
P.E.	1.4	2.6	8.3	16.1	20.4	20.9	19.6	17.6	15.8	11.2	4.3	2.1	140.3
P.	3.0	3.0	1.5	0.8	0.9	7.1	24.7	22.9	15.2	3.7	0.2	0.8	83.8
P.-P.E.	1.6	0.4	-6.8	-15.3	-19.5	-13.8	5.1	5.3	-0.6	-7.5	-4.1	-1.3	-56.5
St.	8.2	8.6	6.6	4.1	1.6	0.9	6.0	11.3	11.0	8.2	6.9	6.6	
Δ St.	1.6	0.4	-2.0	-2.5	-2.5	-0.7	5.1	5.3	-0.3	-2.8	-1.3	-0.3	
A.E.	1.4	2.6	3.5	3.3	3.4	7.8	19.6	17.6	15.5	6.5	1.5	1.1	83.8
W.D.	0.0	0.0	4.8	13.8	17.0	13.1	0.0	0.0	0.3	4.7	2.8	1.0	57.5
W.S.	0.0	0.0	0.0	0.0	0.0	0.0	0.0	0.0	0.0	0.0	0.0	0.0	0.0

Stations	Jan.	Feb.	Mar.	Apr.	May	Jun.	Jul.	Aug.	Sept.	Oct.	Nov.	Dec.	Annual
Roorkee (Moisture Holding Capacity 25.0 cm)													
T°C	13.4	15.8	20.9	26.7	31.5	32.2	29.4	28.7	27.9	24.1	18.3	14.4	
P.E.	1.6	2.6	7.1	14.9	20.2	20.4	19.1	17.4	15.1	9.7	4.0	1.9	134.0
P.	4.3	4.7	2.8	1.1	1.7	9.1	33.7	32.9	20.6	3.9	0.3	1.3	116.4
P.-P.E.	2.7	2.1	-4.3	-13.8	-18.5	-11.3	14.6	15.5	5.5	-5.8	-3.7	-0.6	-17.6
St.	19.4	21.5	18.1	10.4	4.9	3.1	17.7	25.0	25.0	19.7	17.0	16.7	
Δ St.	2.7	2.1	-3.4	-7.7	-5.5	-1.8	14.6	7.3	0.0	-5.3	-2.7	-0.3	
A.E.	1.6	2.6	6.2	8.8	7.2	10.9	19.1	17.4	15.1	9.2	3.0	1.6	102.7
W.D.	0.0	0.0	0.9	6.1	13.0	9.5	0.0	0.0	0.0	0.5	1.0	0.3	31.3
W.S.	0.0	0.0	0.0	0.0	0.0	0.0	0.0	8.2	5.5	0.0	0.0	0.0	13.7
Allahabad (Moisture Holding Capacity 25.0 cm)													
T°C	16.4	19.2	25.2	30.7	34.8	34.4	30.1	29.1	29.0	26.5	21.1	17.1	
P.E.	2.5	4.3	11.7	17.5	21.1	20.9	19.0	17.6	15.9	13.4	6.0	2.7	152.6
P.	2.0	2.2	1.4	0.5	0.8	10.2	27.5	33.3	19.5	4.0	0.7	0.6	102.7
P.-P.E.	-0.5	-2.1	-10.3	-17.0	-20.3	-10.7	8.5	15.7	3.6	-9.4	-5.3	-2.1	-49.9
St.	12.4	11.4	7.6	3.8	1.7	1.1	9.6	25.0	25.0	17.1	13.8	12.7	
Δ St.	-0.3	-1.0	-3.8	-3.8	-2.1	-0.6	8.5	15.4	0.0	-7.9	-3.3	-1.1	
A.E.	2.3	3.2	5.2	4.3	2.9	10.8	19.0	17.6	15.9	11.9	4.0	1.7	98.8
W.D.	0.2	1.1	6.5	13.2	18.2	10.1	0.0	0.0	0.0	1.5	2.0	1.0	53.8
W.S.	0.0	0.0	0.0	0.0	0.0	0.0	0.0	0.3	3.6	0.0	0.0	0.0	3.9

Varanasi (Moisture Holding Capacity 25.0 cm)

Stations	Jan.	Feb.	Mar.	Apr.	May	Jun.	Jul.	Aug.	Sept.	Oct.	Nov.	Dec.	Annual
T°C	16.5	19.3	25.3	30.5	34.3	33.7	30.0	29.1	29.1	26.6	21.0	17.1	
P.E.	2.5	4.3	11.7	17.5	20.8	20.6	19.0	17.5	15.9	13.4	6.0	2.7	151.9
P.	2.5	2.2	1.2	0.4	1.2	9.2	30.6	34.2	22.6	6.0	1.0	0.3	111.4
P.-P.E.	0.0	-2.1	10.5	-17.1	-19.6	-11.4	11.6	16.7	6.7	-7.4	-5.0	-2.4	-40.5
St.	13.7	12.6	8.2	4.1	1.9	1.2	12.8	25.0	25.0	18.5	15.2	13.7	
Δ St.	0.0	-1.1	-4.4	-4.1	-2.2	-0.7	11.6	12.2	0.0	-6.5	-3.3	-1.5	
A.E.	2.5	3.3	5.6	4.5	3.4	9.9	19.0	17.5	15.9	12.5	4.3	1.8	100.2
W.D.	0.0	1.0	6.1	13.0	17.4	10.7	0.0	0.0	0.0	0.9	1.7	0.9	51.7
W.S.	0.0	0.0	0.0	0.0	0.0	0.0	0.0	4.5	6.7	0.0	0.0	0.0	11.2

Gorakhpur (Moisture Holding Capacity 15.0 cm)

Stations	Jan.	Feb.	Mar.	Apr.	May	Jun.	Jul.	Aug.	Sept.	Oct.	Nov.	Dec.	Annual
T°C	16.5	19.1	24.9	29.9	32.5	31.6	29.6	29.3	29.1	26.9	21.6	17.6	
P.E.	2.5	4.0	11.4	17.3	20.2	19.7	18.7	17.6	15.9	13.7	6.2	3.3	150.5
P.	1.9	1.4	1.1	1.1	4.1	18.6	24.2	31.5	25.3	7.4	0.5	0.3	117.4
P.-P.E.	-0.6	-2.6	-10.3	-16.2	-16.1	-1.1	5.5	13.9	9.4	-6.3	-5.7	-3.0	-33.1
St.	5.2	4.4	1.1	0.7	0.2	0.2	5.7	15.0	15.0	9.7	6.6	5.4	
Δ St.	-0.2	-0.8	-3.3	0.4	-0.5	0.0	5.5	9.3	0.0	-5.3	-3.1	-1.2	
A.E.	2.1	2.2	4.4	1.5	4.6	18.6	18.7	17.6	15.9	12.7	3.6	1.5	103.4
W.D.	0.4	1.8	7.0	15.8	15.6	1.1	0.0	0.0	0.0	1.0	2.6	1.8	47.1
W.S.	0.0	0.0	0.0	0.0	0.0	0.0	0.0	4.6	9.4	0.0	0.0	0.0	14.0

Stations	Jan.	Feb.	Mar.	Apr.	May	Jun.	Jul.	Aug.	Sept.	Oct.	Nov.	Dec.	Annual
Gonda (Moisture Holding Capacity 25.0 cm)													
T°C	15.6	18.1	23.7	29.3	32.8	32.2	29.6	29.1	28.7	26.1	20.6	16.6	
P.E.	2.2	3.4	9.6	15.7	20.5	20.0	18.8	17.6	15.6	12.5	5.4	2.4	144.7
P.	2.7	1.6	1.1	0.8	2.7	15.8	37.2	25.3	25.0	6.5	0.2	0.5	129.4
P.-P.E.	0.5	-1.8	-8.5	-15.9	-17.8	-4.2	18.4	17.7	9.4	-6.0	-5.2	-1.9	-15.3
St.	15.2	14.2	10.1	5.3	2.6	2.2	20.6	25.0	25.0	19.6	15.9	14.7	
Δ St.	0.5	-1.0	-4.1	-4.8	-2.7	-0.4	18.4	4.4	0.0	-5.4	-3.7	-1.2	
A.E.	2.2	2.6	5.2	5.6	5.4	15.2	18.8	17.6	15.6	11.9	3.9	1.7	106.7
W.D.	0.0	0.8	4.4	11.1	15.1	3.8	0.0	0.0	0.0	0.6	1.5	0.7	38.0
W.S.	0.0	0.0	0.0	0.0	0.0	0.0	0.0	13.3	9.4	0.0	0.0	0.0	22.7
Bareilly (Moisture Holding Capacity 25.0 cm)													
T°C	15.3	17.9	23.3	29.1	33.2	33.3	30.0	29.1	28.9	25.9	20.5	15.6	
P.E.	1.9	3.4	9.3	16.7	20.7	20.5	18.2	17.6	16.1	12.3	5.4	2.4	144.5
P.	2.5	2.2	1.4	0.8	1.5	10.2	14.6	29.7	17.2	6.2	0.3	0.5	87.1
P.-P.E.	0.6	-1.2	-7.9	-15.9	-19.2	-10.3	-3.6	12.1	1.1	-6.1	-5.1	-1.9	-57.4
St.	8.8	8.4	6.1	3.2	1.5	1.0	0.8	12.9	14.0	10.9	8.9	8.2	
Δ St.	0.6	-0.4	-2.3	-2.9	-1.7	-0.5	-0.2	12.1	1.1	-3.1	-2.0	-0.7	
A.E.	1.9	2.6	3.7	3.7	3.2	10.7	14.8	17.6	16.1	9.3	2.3	1.2	87.1
W.D.	0.0	0.8	5.6	13.0	17.5	9.8	3.4	0.0	0.0	3.0	3.1	1.2	57.4
W.S.	0.0	0.0	0.0	0.0	0.0	0.0	0.0	0.0	0.0	0.0	0.0	0.0	0.0

Stations	Jan.	Feb.	Mar.	Apr.	May	Jun.	Jul.	Aug.	Sept.	Oct.	Nov.	Dec.	Annual
Ballia (Moisture Holding Capacity 25.0 cm)													
T°C	16.9	18.6	23.9	27.8	33.6	33.3	30.1	29.2	29.2	26.6	20.8	17.5	
P.E.	2.8	3.7	9.9	15.7	20.8	20.3	19.0	17.5	15.9	13.4	4.9	3.3	147.2
P.	3.1	0.5	1.7	0.3	2.3	10.0	23.7	25.0	16.1	13.1	1.3	0.3	97.4
P.-P.E.	0.3	-3.2	-8.2	-15.4	-18.5	-10.3	4.7	7.5	0.2	-0.3	-3.6	-3.0	-49.8
St.	10.6	9.3	6.7	3.6	1.7	1.1	5.8	13.3	13.5	13.4	11.5	10.3	
Δ St.	0.3	-1.3	-2.6	-3.1	-1.9	-0.6	4.7	7.5	0.2	-0.1	-1.9	-1.2	
A.E.	2.8	1.8	4.3	3.4	4.2	10.6	19.0	17.5	15.9	13.2	3.2	1.5	97.4
W.D.	0.0	1.9	5.6	12.3	16.6	9.7	0.0	0.0	0.0	0.2	1.7	1.8	49.8
W.S.	0.0	0.0	0.0	0.0	0.0	0.0	0.0	0.0	0.0	0.0	0.0	0.0	0.0
Bahraich (Moisture Holding Capacity 25.0 cm)													
T°C	15.7	18.3	23.7	29.2	32.7	32.3	29.7	29.2	28.9	26.4	21.0	16.9	
P.E.	2.2	3.7	9.6	16.7	20.7	20.1	18.8	17.6	16.0	13.0	5.9	2.7	147.0
P.	2.8	1.7	1.2	0.7	2.3	14.2	26.1	28.5	21.0	8.3	0.2	0.7	107.7
P.-P.E.	0.6	-2.0	-8.4	-16.0	-18.4	-5.9	7.3	10.9	5.0	-4.7	-5.7	-2.0	-39.3
St.	15.8	14.0	10.4	5.4	2.6	2.0	9.3	20.2	25.0	20.7	16.5	15.2	
Δ St.	0.6	-1.8	-3.6	-5.0	-2.8	-0.6	7.3	10.9	4.8	-4.3	-4.2	-1.3	
A.E.	2.2	3.5	4.8	5.7	5.1	14.8	18.8	17.6	16.0	12.6	4.4	2.0	107.5
W.D.	0.0	0.2	4.8	11.0	15.6	5.3	0.0	0.0	0.0	0.4	1.5	0.7	39.5
W.S.	0.0	0.0	0.0	0.0	0.0	0.0	0.0	0.0	0.2	0.0	0.0	0.0	0.2

Stations	Jan.	Feb.	Mar.	Apr.	May	Jun.	Jul.	Aug.	Sept.	Oct.	Nov.	Dec.	Annual
Kheri (Moisture Holding Capacity 25.0 cm)													
T°C	15.3	18.5	23.7	29.2	32.8	32.2	29.3	29.0	28.8	25.9	20.8	17.0	
P.E.	1.9	3.7	9.6	16.7	20.7	20.2	19.5	18.3	15.8	12.3	5.7	2.7	147.1
P.	4.4	1.2	1.9	0.2	2.4	13.5	32.9	30.5	15.8	10.1	0.1	0.2	113.2
P.-P.E.	2.5	-2.5	-7.7	-16.5	-18.3	-6.7	13.4	12.2	0.0	-2.2	-5.6	-2.5	-33.9
St.	19.0	17.2	12.6	6.5	3.1	2.4	15.8	25.0	25.0	22.9	18.2	16.5	
Δ St.	2.5	-1.8	-4.6	-6.1	-3.4	-0.7	13.4	9.2	0.0	-2.1	-4.7	-1.7	
A.E.	1.9	3.0	6.5	6.3	5.8	14.2	19.5	18.3	15.8	12.2	4.8	1.9	110.2
W.D.	0.0	0.7	3.1	10.4	14.9	6.0	0.0	0.0	0.0	0.1	0.9	0.8	36.9
W.S.	0.0	0.0	0.0	0.0	0.0	0.0	0.0	3.0	0.0	0.0	0.0	0.0	3.0
Aligarh (Moisture Holding Capacity 25.0 cm)													
T°C	14.7	17.4	22.8	28.9	33.8	34.2	30.8	29.4	29.0	26.2	20.6	16.0	
P.E.	1.6	3.1	8.7	16.8	21.2	–21.1	20.2	18.1	16.1	12.6	5.3	2.4	147.2
P.	1.8	1.5	1.4	1.1	1.2	6.8	22.7	21.6	15.3	3.6	0.2	0.2	77.4
P.-P.E.	0.2	-1.6	-7.3	-15.7	-20.0	-14.3	2.5	3.5	-0.8	-9.0	-5.1	-2.2	-69.8
St.	3.4	3.2	2.4	1.3	0.6	0.3	2.8	6.3	6.1	4.2	3.5	3.2	
Δ St.	0.2	-0.2	-0.8	-1.1	-0.7	-0.3	2.5	3.5	-0.2	-1.9	-0.7	-0.3	
A.E.	1.6	1.7	2.2	2.2	1.9	7.1	20.2	18.1	15.5	5.5	0.9	0.5	77.4
W.D.	0.0	1.4	6.5	14.6	19.3	14.0	0.0	0.0	0.6	7.1	4.4	1.9	69.8
W.S.	0.0	0.0	0.0	0.0	0.0	0.0	0.0	0.0	0.0	0.0	0.0	0.0	0.0

Stations	Jan.	Feb.	Mar.	Apr.	May	Jun.	Jul.	Aug.	Sept.	Oct.	Nov.	Dec.	Annual
Mairpuri (Moisture Holding Capacity 25.0 cm)													
T°C	15.2	18.1	23.7	29.7	34.2	34.8	30.9	29.5	29.2	26.5	20.9	16.5	
P.E.	1.9	3.4	9.6	17.0	20.9	21.1	19.8	18.0	16.0	13.3	5.9	3.8	150.7
P.	1.7	0.9	1.2	0.6	1.1	6.4	19.6	24.7	15.2	4.0	0.2	0.6	76.2
P.-P.E.	-0.2	-2.5	-8.4	-16.4	-19.8	-14.7	-0.2	6.7	-0.8	-9.3	-5.7	-3.2	-74.5
St.	3.2	2.9	2.0	1.1	0.5	0.2	0.2	6.9	6.7	4.6	3.7	3.2	
Δ St.	0.0	-0.3	-0.9	-0.9	-0.6	-0.3	0.0	6.7	-0.2	-2.1	-0.9	-0.5	
A.E.	1.7	1.2	2.1	1.5	1.7	6.7	19.6	18.0	15.4	6.1	1.1	1.1	76.2
W.D.	0.2	2.2	7.5	15.5	19.2	14.4	0.2	0.0	0.6	7.2	4.8	2.7	74.5
W.S.	0.0	0.0	0.0	0.0	0.0	0.0	0.0	0.0	0.0	0.0	0.0	0.0	0.0
Banda (Moisture Holding Capacity 10.0 cm)													
T°C	16.6	19.8	25.7	31.1	35.5	35.1	30.2	29.8	28.9	26.6	21.0	17.4	
P.E.	2.5	4.5	12.7	17.9	21.1	20.9	19.0	18.1	15.9	13.4	6.0	3.3	155.3
P.	2.3	1.3	1.0	0.2	0.7	7.2	28.0	38.0	14.2	7.0	0.4	0.2	100.5
P.-P.E.	-0.2	-3.2	-11.7	-17.7	-20.4	-13.7	9.0	19.9	-1.7	-6.4	-5.6	-3.1	54.8
St.	1.8	1.2	0.4	0.1	0.0	0.0	9.0	10.0	8.4	4.4	2.4	1.9	
Δ St.	-0.1	-0.6	-0.8	-0.3	-0.1	0.0	9.0	1.0	-1.6	-4.0	-2.0	-0.5	
A.E.	2.4	1.9	1.8	0.5	0.8	7.2	19.0	18.1	15.8	11.0	2.4	0.7	81.6
W.D.	0.1	2.6	10.9	17.4	20.3	13.7	0.0	0.0	0.1	2.4	3.6	2.6	73.7
W.S.	0.0	0.0	0.0	0.0	0.0	0.0	0.0	18.9	0.0	0.0	0.0	0.0	18.9

Stations	*Jan.*	*Feb.*	*Mar.*	*Apr.*	*May*	*Jun.*	*Jul.*	*Aug.*	*Sept.*	*Oct.*	*Nov.*	*Dec.*	*Annual*
Lucknow (Moisture Holding Capacity 25.0 cm)													
T°C	16.1	18.9	24.6	30.1	33.8	33.7	30.1	29.3	29.1	26.3	20.5	17.0	
P.E.	2.5	4.0	10.5	17.3	20.9	20.7	19.1	17.6	15.9	12.8	5.4	2.7	149.4
P.	2.4	1.7	0.9	0.6	1.2	9.4	29.9	30.2	18.2	4.0	0.1	0.6	99.2
P.-P.E.	-0.1	-2.3	-9.6	-16.7	-19.7	-11.3	10.8	12.6	2.3	-8.8	-5.3	-2.1	-50.2
St.	12.9	11.8	8.0	4.1	1.8	1.2	12.0	24.6	25.0	17.5	14.2	13.0	
Δ St.	-0.1	-1.1	-3.8	-3.9	-2.3	-0.6	10.8	12.6	0.4	-7.5	-3.3	-1.2	
A.E.	2.5	2.8	4.7	4.5	3.5	10.0	19.1	17.6	15.9	11.5	3.4	1.8	97.3
W.D.	0.0	1.2	5.8	12.8	17.4	10.7	0.0	0.0	0.0	1.3	2.0	0.9	52.1
W.S.	0.0	0.0	0.0	0.0	0.0	0.0	0.0	0.0	1.9	0.0	0.0	0.0	1.9

T°C = Mean monthly temperature;
P. = Precipitation;
St. = Soil moisture storage;
A.E. = Actual evapotranspiration;
W.S. = Water surplus.
P.E. = Potential evapotranspiration;
P.-P.E. = Precipitation — Potential evapotranspiration;
Δ St. = Change in soil moisture storage;
W.D. = Water deficit;

Source: Calculated by the authoress on the basis of average data of Temperature and Rainfall (The calculation made by the authoress are on the basis of the modus operendi explained in the book.

APPENDIX F

Average Annual Water Surplus, Water Deficit and Moisture Indices

Stations	*W.S. (cm)*	*W.D. (cm)*	I_h	I_a	I_m	I_{ma}
Mussoorie	170.6	6.3	234.66	8.67	225.99	91.33
Dehra dun	136.5	24.3	114.51	20.38	94.13	79.61
Mukteswar	68.8	5.1	95.16	7.05	88.11	92.95
Nainital	190.7	3.5	262.67	4.82	257.85	95.18
Joshimath	62.0	6.3	78.88	8.02	70.86	91.98
Tehri	33.9	35.8	27.67	29.14	-1.47	70.86
Mukhim	96.3	3.8	122.99	4.84	118.14	95.15
Munsiari	199.3	0.3	286.76	0.43	286.33	99.57
Roorkee	13.7	31.3	10.22	23.36	-13.14	76.64
Najibabad	23.6	33.2	17.49	24.61	-7.12	75.39
Pantnagar	38.1	23.3	30.41	18.60	11.81	81.40
Bahraich	0.2	39.5	0.14	26.87	-26.73	73.13
Nautanwa	53.6	34.7	37.67	24.39	13.28	75.61
Meerut	0.0	57.5	0.00	40.98	-40.98	59.73
Aligarh	0.0	69.8	0.00	47.42	-47.42	52.58
Agra (Obsy.)	0.0	71.3	0.00	48.21	-48.21	51.86
Agra	0.0	79.8	0.00	53.99	-53.99	40.01
Mainpuri	0.0	74.5	0.00	49.34	-49.34	50.56
Jhansi	9.4	66.2	6.16	40.73	-34.57	59.27
Orai	19.0	65.1	12.62	43.22	-33.93	56.77
Banda	18.9	73.7	12.17	47.46	-35.29	52.54
Bareilly	0.0	57.4	0.00	39.79	-39.79	59.82
Kheri	3.0	36.9	2.04	25.08	-23.04	74.92
Hardoi	8.5	43.4	5.63	28.72	-23.09	73.09
Lucknow (Obsy.)	1.9	52.1	1.27	34.87	-33.60	65.13
Lucknow	4.7	51.5	3.16	34.59	-31.43	65.41
Kanpur (Obsy.)	0.0	60.1	0.00	40.53	-40.53	59.47
Kanpur	0.0	60.1	0.00	40.53	-40.53	59.50
Fatehpur	0.0	63.2	0.00	41.70	-41.70	58.34
Allahabad	3.9	53.8	2.56	35.26	-32.70	64.74
Gonda	22.7	38.0	15.69	26.26	-10.57	73.74
Gorakhpur	14.0	47.1	9.30	31.30	-22.00	68.70
Azamgarh	6.9	54.2	4.54	35.63	-31.09	64.37
Varanasi (Obsy.)	11.2	51.7	8.68	34.03	-25.35	65.96
Varanasi	8.2	52.5	5.39	34.56	-29.17	65.44
Faizabad	15.3	43.6	10.64	30.32	-19.68	69.68
Sultanpur	8.3	47.0	5.57	31.56	-25.99	68.44
Ballia	0.0	49.8	0.00	33.83	-33.83	66.17
Pipri	0.0	64.7	0.00	37.95	-37.95	62.05

W.S. = Water surplus; W.D. = Water deficit; I_h = Humidity Index;
I_a = aridity index: I_m = Mositure index; I_{ma} = Moisture adequacy index.
Source: Calculated by the authoress on the basis of Thornthwaite's Method (The calculations made by the authoress are on the basis of the modus operendi explained in the book).

APPENDIX G

Climatic Water Balance Tables (Penman)

	Jan.	Feb.	Mar.	Apr.	May	Jun.	Jul.	Aug.	Sept.	Oct.	Nov.	Dec.	Annual
Dehra Dun (Moisture Holding Capacity 5.0 cm)													
P.E.	3.79	5.56	9.66	13.42	16.89	16.21	11.23	10.23	10.35	9.15	5.22	3.53	115.24
P.	6.20	5.90	4.00	1.70	4.20	18.60	72.10	76.50	32.70	7.10	0.50	1.90	231.40
P.-P.E.	2.41	0.34	-5.66	-11.72	-12.69	2.39	60.87	66.27	22.35	-2.05	-4.72	-1.63	116.16
St.	3.21	3.55	1.00	0.1-	0.00	2.39	5.00	5.00	5.00	3.20	1.10	0.80	
ΔSt.	2.41	0.34	-2.55	-0.90	-0.10	2.39	2.61	0.00	0.00	-1.80	-2.10	-0.30	
A.E.	3.79	5.56	6.55	2.60	4.30	16.21	11.23	10.23	10.35	8.90	2.60	2.20	84.52
W.D.	0.00	0.00	3.11	10.82	12.59	0.00	0.00	0.00	0.00	0.25	2.62	1.33	30.72
W.S.	0.00	0.00	0.00	0.00	0.00	0.00	58.26	66.27	22.35	0.00	0.00	0.00	146.88
Roorkee (Moisture Holding Capacity 25.0 cm)													
P.E.	4.28	6.24	11.04	15.27	19.89	19.20	13.53	12.38	12.16	9.94	5.55	3.85	133.33
P.	4.30	4.70	2.80	1.10	1.70	9.10	33.70	32.90	20.60	3.90	0.30	1.30	116.40
P.-P.E.	0.02	-1.54	-8.24	-14.17	-18.19	-10.10	20.17	20.52	8.44	-6.04	-5.25	-2.55	-16.93
St.	14.32	13.40	9.70	5.40	2.60	1.70	21.87	25.00	25.00	19.60	15.90	14.30	
ΔSt.	0.02	-0.92	-3.70	-4.30	-2.80	-0.90	20.17	3.13	0.00	-5.40	-3.70	-1.60	
A.E.	4.28	5.62	6.50	5.40	4.50	10.00	13.53	12.38	12.16	9.30	4.00	2.90	90.57
W.D.	0.00	0.62	4.54	9.87	15.39	9.20	0.00	0.00	0.00	0.64	1.55	0.95	42.76
W.S.	0.00	0.00	0.00	0.00	0.00	0.00	0.00	17.39	8.44	0.00	0.00	0.00	25.83

	Jan.	Feb.	Mar.	Apr.	May	Jun.	Jul.	Aug.	Sept.	Oct.	Nov.	Dec.	Annual
Gonda (Moisture Holding Capacity 25.0 cm)													
P.E.	5.18	7.48	13.46	17.70	20.65	18.33	12.84	11.65	11.95	11.28	6.88	4.74	142.23
P.	2.70	1.60	1.10	0.80	2.70	15.80	37.20	35.30	25.00	6.50	0.20	0.50	129.40
P.-P.E.	-2.48	-5.88	-12.36	-16.99	-17.95	-2.53	24.36	23.65	13.05	-4.78	-6.68	-4.24	-13.17
St.	12.00	9.50	5.70	2.90	1.40	1.30	25.00	25.00	25.00	20.60	15.70	13.20	
ΔSt.	-1.20	-2.50	-3.80	-2.80	-1.50	-0.10	23.70	0.00	0.00	-4.40	-4.90	-2.50	
A.E.	3.90	4.10	4.90	3.60	4.20	15.90	12.84	11.65	11.95	10.90	5.10	3.00	92.04
W.D.	1.28	3.38	8.56	14.19	16.45	2.43	0.00	0.00	0.00	0.38	1.78	1.74	50.19
W.S.	0.00	0.00	0.00	0.00	0.00	0.00	0.66	23.65	13.05	0.00	0.00	0.00	37.36
Jhansi (Moisture Holding Capacity 20.0 cm)													
P.E.	6.35	8.49	13.83	17.23	21.51	20.11	13.65	11.63	12.70	12.28	7.91	5.86	151.55
P.	1.70	1.40	1.10	0.30	0.80	10.50	29.80	30.90	19.00	3.20	0.50	0.70	99.90
P.-P.E.	-4.65	-7.09	-12.73	-16.92	-20.71	-9.61	16.15	19.27	6.30	-9.08	-7.41	-5.16	-51.65
St.	5.30	3.70	1.90	0.70	0.30	0.20	16.35	20.00	20.00	12.60	8.70	6.70	
ΔSt.	-1.40	-1.60	-1.80	-1.20	-0.40	-0.10	16.15	3.65	0.00	-7.40	-3.90	-2.00	
A.E.	3.10	3.00	2.90	1.50	1.20	10.60	13.65	11.63	12.70	10.60	4.40	2.70	77.98
W.D.	3.25	5.49	10.93	15.73	20.31	9.51	0.00	0.00	0.00	1.68	3.51	3.16	73.57
W.S.	0.00	0.00	0.00	0.00	0.00	0.00	0.00	15.62	6.30	0.00	0.00	0.00	21.92

	Jan.	Feb.	Mar.	Apr.	May	Jun.	Jul.	Aug.	Sept.	Oct.	Nov.	Dec.	Annual
Meerut (Moisture Holding Capacity 25.0 cm)													
P.E.	5.31	7.51	12.71	17.47	22.22	22.53	16.30	14.21	14.22	11.13	6.59	4.33	154.53
P.	3.00	3.00	1.50	0.80	0.90	7.10	24.70	22.90	15.20	3.70	0.20	0.80	83.80
P.-P.E.	-2.31	-4.51	-11.21	-16.67	-21.32	-15.43	8.40	8.69	0.98	-7.43	-6.39	-3.53	-70.73
St.	8.40	7.00	4.40	2.30	1.00	0.50	8.90	17.59	18.57	13.80	10.70	9.20	
ΔSt.	-0.80	-1.40	-2.60	-2.10	-1.30	-0.50	8.40	8.69	0.98	-4.77	-3.10	-1.50	
A.E.	3.80	4.40	4.10	2.90	2.20	7.60	16.30	14.21	14.22	8.47	3.30	2.30	83.80
W.D.	1.51	3.11	8.61	14.57	20.02	14.93	0.00	0.00	0.00	2.66	3.29	2.03	70.73
W.S.	0.00	0.00	0.00	0.00	0.00	0.00	0.00	0.00	0.00	0.00	0.00	0.00	0.00
Kheri (Moisture Holding Capacity 25.0 cm)													
P.E.	4.67	7.12	12.09	16.75	19.1	18.19	12.80	11.98	12.32	10.93	6.51	4.38	136.91
P.	4.40	1.20	1.90	0.20	2.40	13.50	32.90	30.50	15.80	10.10	0.10	0.20	113.20
P.-P.E.	-0.27	-5.92	10.19	-16.55	-16.77	-4.69	20.10	18.52	-3.42	-0.83	-6.41	-4.18	-23.71
St.	15.60	12.30	8.10	4.10	2.10	1.70	21.80	25.00	25.00	24.20	18.70	15.80	
ΔSt.	-0.20	-3.30	-4.20	-4.00	-2.00	-0.40	20.10	3.20	0.00	-0.80	-5.50	-2.90	
A.E.	4.60	4.50	6.10	4.20	4.40	13.90	12.80	11.98	12.32	10.90	5.60	3.10	94.40
W.D.	0.07	2.62	5.99	12.55	14.77	4.29	0.00	0.00	0.00	0.03	0.91	1.28	42.51
W.S.	0.00	0.00	0.00	0.00	0.00	0.00	0.00	15.32	3.48	0.00	0.00	0.00	18.80

	Jan.	Feb.	Mar.	Apr.	May	Jun.	Jul.	Aug.	Sept.	Oct.	Nov.	Dec.	Annual
Bahraich (Moisture Holding Capacity 25.0 cm)													
P.E.	5.11	7.34	12.92	17.64	21.26	18.92	13.24	11.79	12.08	11.46	6.95	4.78	143.49
P.	2.80	1.70	1.20	0.70	2.30	14.20	26.10	28.50	21.00	8.30	0.20	0.70	107.70
P.-P.E.	-2.31	-5.64	-11.72	-16.94	-18.96	-4.72	12.86	16.71	8.92	-3.16	-6.75	-4.08	-35.79
St.	12.90	10.30	6.40	3.20	1.50	1.20	14.06	25.00	25.00	22.00	16.80	14.20	
ΔSt.	-1.30	-2.60	-3.90	-3.20	-1.70	-0.30	12.86	10.94	0.00	-3.00	-5.20	-2.60	
A.E.	4.10	4.30	5.10	3.90	4.00	14.50	13.24	11.79	12.08	11.30	5.40	3.30	93.01
W.D.	1.01	3.04	7.82	13.74	17.25	4.42	0.00	0.00	0.00	0.16	1.55	1.48	50.48
W.S.	0.00	0.00	0.00	0.00	0.00	0.00	0.00	5.77	8.92	0.00	0.00	0.00	14.69
Aligarh (Moisture Holding Capacity 25.0 cm)													
P.E.	5.38	7.62	12.98	17.45	22.18	21.47	15.11	12.71	13.63	11.94	7.29	5.15	152.91
P.	1.80	1.50	1.40	1.10	1.20	6.80	22.70	21.60	15.30	3.60	0.20	0.20	77.40
P.-P.E.	-3.58	-6.12	-11.58	-16.35	-20.98	-14.67	7.59	8.89	1.67	08.34	-7.09	-4.95	-75.51
St.	7.00	5.50	3.40	1.80	0.80	0.40	7.99	16.88	18.55	13.20	9.90	8.10	
ΔSt.	-1.10	-1.50	-2.10	-1.60	-1.00	-0.40	7.59	8.89	1.67	05.35	-3.30	-1.80	
A.E.	2.90	3.00	3.50	2.70	2.20	7.20	15.11	12.71	13.63	8.95	3.50	2.00	77.40
W.D.	2.48	4.62	9.48	14.75	19.98	14.27	0.00	0.00	0.00	2.99	3.79	3.15	75.51
W.S.	0.00	0.00	0.00	0.00	0.00	0.00	0.00	0.00	0.00	0.00	0.00	0.00	0.00

	Jan.	Feb.	Mar.	Apr.	May	Jun.	Jul.	Aug.	Sept.	Oct.	Nov.	Dec.	Annual
Mainpuri (Moisture Holding Capacity 25.0 cm)													
P.E.	5.02	7.25	12.51	16.50	20.91	19.48	13.58	12.11	12.63	11.49	6.87	4.77	143.12
P.	1.70	0.90	1.20	0.60	1.10	6.40	19.60	24.70	15.20	4.00	0.20	0.60	76.20
P.-P.E.	-3.32	-6.35	-11.31	-15.90	-19.81	-13.08	6.02	12.59	2.57	-7.49	-6.67	-4.17	-66.92
St.	9.10	7.00	4.50	2.40	1.10	0.60	6.62	19.21	21.78	16.20	12.30	10.50	
ΔSt.	-1.40	-2.10	-2.50	-2.10	-1.30	-0.50	6.02	12.59	2.57	-5.58	-3.90	-1.80	
A.E.	3.10	3.00	3.70	2.70	2.40	6.90	13.58	12.11	12.63	9.58	4.10	2.40	76.20
W.D.	1.92	4.25	8.81	13.80	18.51	12.58	0.00	0.00	0.00	1.91	2.77	2.37	66.92
W.S.	0.00	0.00	0.00	0.00	0.00	0.00	0.00	0.00	0.00	0.00	0.00	0.00	0.00
Banda (Moisture Holding Capacity 10.0 cm)													
P.E.	5.44	7.79	13.02	16.72	21.03	19.51	13.68	12.10	12.80	11.50	7.20	5.01	145.62
P.	2.30	1.30	1.00	0.20	0.70	7.20	28.00	38.00	14.20	7.00	0.40	0.20	100.50
P.-P.E.	-3.14	-6.49	-12.02	-16.52	-20.33	-12.31	14.32	25.90	1.40	-4.50	-6.62	-4.81	-45.12
St.	1.40	0.70	0.20	0.00	0.00	0.00	10.00	10.00	10.00	6.30	3.20	2.00	
ΔSt.	-0.60	-0.70	-0.50	-0.20	0.00	0.00	10.00	0.00	0.00	-3.70	-3.10	1.20	
A.E.	2.90	2.00	1.50	0.40	0.70	7.20	13.68	12.10	12.80	10.70	3.50	1.40	68.88
W.D.	2.54	5.79	11.52	16.32	20.33	12.31	0.00	0.00	0.00	0.80	3.52	3.61	76.74
W.S.	0.00	0.00	0.00	0.00	0.00	0.00	4.32	25.90	1.40	0.00	0.00	0.00	31.62

	Jan.	Feb.	Mar.	Apr.	May	Jun.	Jul.	Aug.	Sept.	Oct.	Nov.	Dec.	Annual
Lucknow (Moisture Holding Capacity 25.0 cm)													
P.E.	6.68	8.20	14.17	18.99	22.97	20.95	13.15	11.35	12.16	11.35	7.07	4.86	151.90
P.	2.40	1.70	0.90	0.60	1.20	9.40	29.90	30.20	18.20	4.00	0.10	0.60	99.20
P.-P.E.	-4.28	-6.50	-13.27	-18.39	-21.77	-11.55	16.75	18.85	6.04	-7.35	-6.97	-4.26	-52.68
St.	9.90	7.60	4.50	2.10	0.90	0.60	17.35	25.00	25.00	18.50	14.00	11.80	
ΔSt.	-1.90	-2.30	-3.10	-2.40	-1.20	-0.30	16.75	7.65	0.00	-6.50	-4.50	-2.20	
A.E.	4.30	4.00	4.00	3.00	2.40	9.70	13.15	11.35	12.16	10.50	4.60	2.80	81.96
W.D.	2.38	4.20	10.17	15.99	20.57	11.25	0.00	0.00	0.00	0.85	2.47	2.06	69.94
W.S.	0.00	0.00	0.00	0.00	0.00	0.00	0.00	11.20	6.04	0.00	0.00	0.00	17.24
Kanpur (Moisture Holding Capacity 25.0 cm)													
P.E.	6.07	8.88	15.73	20.53	25.38	22.98	14.66	12.54	13.28	12.55	7.83	5.60	166.03
P.	2.30	1.60	0.90	0.50	0.60	6.80	20.80	28.60	20.20	4.30	0.80	0.80	88.20
P.-P.E.	-3.77	-7.28	-14.83	-20.03	-24.78	-16.18	6.14	16.06	6.92	-8.25	-7.03	-4.80	-77.83
St.	9.60	7.10	3.90	1.70	0.60	0.30	6.44	22.50	25.00	17.90	13.50	11.10	
ΔSt.	-1.50	-2.50	-3.20	-2.20	-1.10	-0.30	6.14	16.06	2.50	-7.10	-4.40	-2.40	
A.E.	3.80	4.10	4.10	2.70	1.70	7.10	14.66	12.54	13.28	11.40	5.20	3.20	83.78
W.D.	2.27	4.78	11.63	17.83	23.58	15.88	0.00	0.00	0.00	1.15	2.63	2.40	82.25
W.S.	0.00	0.00	0.00	0.00	0.00	0.00	0.00	0.00	4.42	0.00	0.00	0.00	4.42

	Jan.	Feb.	Mar.	Apr.	May	Jun.	Jul.	Aug.	Sept.	Oct.	Nov.	Dec.	Annual
Allahabad (Moisture Holding Capacity 25.0 cm)													
P.E.	6.19	8.60	14.71	18.78	23.34	20.58	12.77	11.03	12.01	12.16	7.82	-5.70	153.69
P.	2.00	2.20	1.40	0.50	0.80	10.20	27.50	33.30	19.50	4.00	0.70	0.60	102.70
P.-P.E.	-4.19	6.40	-13.31	-18.28	-22.54	-10.38	14.73	22.27	7.49	-8.16	-7.12	-5.10	-70.98
St.	9.30	7.10	4.20	2.00	0.80	0.5	15.23	25.00	25.00	17.90	13.50	10.90	
ΔSt.	-1.60	-2.20	-2.90	-2.20	-1.20	-0.30	14.73	9.77	0.00	-7.10	-4.40	-2.60	
A.E.	3.60	4.40	4.30	2.70	2.00	10.50	12.77	11.03	12.01	11.10	5.10	3.20	82.71
W.D.	2.59	4.20	10.41	16.08	21.34	10.08	0.00	0.00	0.00	1.06	2.72	2.50	70.98
W.S.	0.00	0.00	0.00	0.00	0.00	0.00	0.00	12.50	7.40	0.00	0.00	0.00	19.99
Varanasi (Moisture Holding Capacity 25.0 cm)													
P.E.	6.11	9.55	15.73	21.03	24.91	20.44	13.19	11.67	12.13	12.32	8.01	5.73	160.82
P.	2.50	2.20	1.20	0.40	1.20	9.20	30.60	34.20	22.60	6.00	1.00	0.30	111.40
P.-P.E.	-3.61	-7.35	-14.53	-20.63	-23.71	-11.24	17.41	22.53	10.47	-6.32	-7.01	-5.43	-49.42
St.	10.10	7.60	4.20	1.80	0.70	0.50	17.91	25.00	25.00	19.30	14.60	11.70	
ΔSt.	-1.60	-.250	-3.40	-2.40	-1.10	-0.20	17.41	7.09	0.00	-5.70	-4.70	-2.90	
A.E.	4.10	4.70	4.60	2.80	2.30	9.40	13.19	11.67	12.13	11.70	5.70	3.20	85.49
W.D.	2.01	4.85	11.13	18.23	22.61	11.04	0.00	0.00	0.00	0.62	2.31	2.53	75.33
W.S.	0.00	0.00	0.00	0.00	0.00	0.00	0.00	15.44	10.47	0.00	0.00	0.00	25.91

	Jan.	Feb.	Mar.	Apr.	May	Jun.	Jul.	Aug.	Sept.	Oct.	Nov.	Dec.	Annual
Gorakhpur (Moisture Holding Capacity 15.0 cm)													
P.E.	5.35	7.65	13.55	17.72	20.72	17.11	13.06	11.93	11.61	11.47	7.18	5.02	142.37
P.	1.90	1.40	1.10	1.10	4.10	18.60	24.20	31.50	25.30	7.40	0.50	0.30	117.40
P.-P.E.	-3.45	-6.25	-12.45	-16.62	-16.62	1.49	11.14	19.57	13.69	-4.07	-6.68	-4.72	-24.97
St.	4.10	2.70	1.20	0.40	0.10	1.59	12.73	15.00	15.00	11.30	7.20	5.20	
ΔSt.	-1.10	-1.40	-1.50	-0.80	-0.30	1.49	11.14	2.27	0.00	-3.70	-4.10	-2.00	
A.E.	3.00	2.80	2.60	1.90	4.40	17.11	13.06	11.93	11.61	11.10	4.60	2.30	86.41
W.D.	2.35	4.85	10.95	15.82	16.32	0.00	0.00	0.00	0.00	0.37	2.58	2.72	55.96
W.S.	0.00	0.00	0.00	0.00	0.00	0.00	0.00	17.30	13.69	0.00	0.00	0.00	30.99
Agra (Moisture Holding Capacity 20.0 cm)													
P.E.	5.29	7.52	12.64	16.26	20.83	20.44	14.58	12.39	13.29	11.61	6.87	4.97	146.69
P.	1.60	0.90	1.10	0.50	1.00	6.00	21.00	26.30	15.20	2.40	0.20	0.40	76.60
P.-P.E.	-3.69	-6.62	-11.54	-15.76	-19.83	-14.44	6.42	13.91	1.91	-9.21	-6.67	-4.57	-70.09
St.	5.90	4.20	2.40	1.10	0.40	0.20	6.62	20.00	20.00	12.50	8.90	7.10	
ΔSt.	-1.20	-1.70	-1.80	-1.30	-0.70	-0.20	6.42	13.38	0.00	-7.50	-3.60	-1.80	
A.E.	2.80	2.60	2.90	1.80	1.70	6.20	14.58	12.39	13.29	9.90	3.80	2.20	74.16
W.D.	2.49	4.92	9.74	14.46	19.13	14.24	0.00	0.00	0.00	1.71	3.07	2.77	72.53
W.S.	0.00	0.00	0.00	0.00	0.00	0.00	0.00	0.53	1.91	0.00	0.00	0.00	2.44

	Jan.	Feb.	Mar.	Apr.	May	Jun.	Jul.	Aug.	Sept.	Oct.	Nov.	Dec.	Annual
Bareilly (Moisture Holding Capacity 25.0 cm)													
P.E.	4.88	7.06	12.47	16.65	20.85	19.60	12.94	11.55	12.26	11.01	6.42	4.53	140.22
P.	2.50	2.20	1.40	0.80	1.50	10.20	14.60	29.70	17.20	6.20	0.30	0.50	87.10
P.-P.E.	-2.38	-4.86	-11.07	-15.85	-19.35	-9.40	1.66	18.15	4.94	-4.81	-6.12	-4.03	-53.12
St.	12.40	10.20	6.50	3.40	1.60	1.10	2.76	20.91	25.00	20.60	16.10	13.60	
ΔSt.	-1.20	-2.20	-3.70	-3.10	-1.80	-0.50	1.66	18.15	4.09	-4.40	-4.50	-2.50	
A.E.	3.70	4.40	5.10	3.90	3.30	10.70	12.94	11.55	12.26	10.60	4.80	3.00	86.25
W.D.	1.18	2.66	7.37	12.75	17.55	8.90	0.00	0.00	0.00	4.10	1.60	1.53	53.97
W.S.	0.00	0.00	0.00	0.00	0.00	0.00	0.00	0.00	0.85	0.00	0.00	0.00	0.85

Source: Calculated by the authoress on the basis on Penman's P.E. (The calculations made by the authoress are on the basis of the modus operendi explained in the book).

APPENDIX H

Water Balance of Excessive Rainy Year

	Jan.	Feb.	Mar.	Apr.	May	Jun.	Jul.	Aug.	Sept.	Oct.	Nov.	Dec.	Annual
Mussoorie (Moisture Holding Capacity 5.0 cm) — 1971													
T°C	6.1	8.4	13.1	16.2	16.7	18.1	17.8	17.3	16.9	15.3	11.0	10.2	
P.E.	1.6	2.3	5.3	7.1	8.1	9.1	9.4	8.6	7.4	6.2	3.5	3.2	71.8
P.	4.2	1.5	0.0	7.4	24.7	46.8	80.9	85.0	27.4	5.9	6.1	0.0	289.9
P.-P.E.	2.6	-0.8	-5.3	0.3	16.6	37.7	71.5	76.4	20.0	-0.3	2.6	-3.2	218.1
St.	5.0	4.2	1.4	1.7	5.0	5.0	5.0	5.0	5.0	4.7	5.0	2.5	
ΔSt.	2.5	-0.8	-2.8	0.3	3.3	0.0	0.0	0.0	0.0	-0.3	0.3	-2.5	
A.E.	1.6	2.3	2.8	7.1	8.1	9.1	9.4	8.6	7.4	6.2	3.5	2.5	68.6
W.D.	0.0	0.0	2.5	0.0	0.0	0.0	0.0	0.0	0.0	0.0	0.0	0.7	3.2
W.S.	0.1	0.0	0.0	0.0	13.3	37.7	37.5	76.4	20.0	0.0	2.3	0.0	221.3
Dehra Dun (Moisture Holding Capacity 5.0 cm) — 1971													
T°C	11.6	14.2	19.0	24.0	25.1	25.1	24.9	23.9	25.2	22.5	17.9	14.9	
P.E.	1.6	2.6	6.2	11.3	13.8	13.7	14.0	13.0	12.4	9.1	4.8	2.9	105.4
P.	4.0	4.3	0.5	3.1	7.2	79.6	51.7	104.4	43.3	10.1	3.5	0.0	311.7
P.-P.E.	2.4	1.7	-5.7	-8.2	-6.6	65.9	37.7	91.4	30.9	1.0	-1.3	-2.9	206.3
St.	4.4	5.0	1.5	0.3	0.1	5.0	5.0	5.0	5.0	5.0	3.8	2.0	
ΔSt.	2.4	0.6	-3.5	-1.2	-0.2	4.9	0.0	0.0	0.0	0.0	-1.2	-1.8	
A.E.	1.6	2.6	4.0	4.3	7.4	13.7	14.0	13.0	12.4	9.1	4.7	1.8	88.6
W.D.	0.0	0.0	2.2	7.0	6.4	0.0	0.0	0.0	0.0	0.0	0.1	1.1	16.8
W.S.	0.0	1.1	0.0	0.0	0.0	61.0	37.7	91.4	30.9	1.0	0.0	0.0	223.1

	Jan.	Feb.	Mar.	Apr.	May	Jun.	Jul.	Aug.	Sept.	Oct.	Nov.	Dec.	Annual
Mukteswar (Moisture Holding Capacity 5.0 cm) — 1971													
T°C	5.8	7.8	11.9	15.0	15.0	16.9	17.0	16.1	16.1	13.3	10.7	9.8	
P.E.	1.6	2.3	4.9	6.8	7.4	8.7	9.0	7.8	7.1	5.3	3.7	3.2	67.8
P.	4.6	4.9	4.9	3.7	13.8	22.8	43.1	35.2	33.2	15.9	1.8	0.1	184.0
P.-P.E.	3.0	2.6	0.0	-3.1	6.4	14.1	34.1	27.4	26.1	10.6	-1.9	-3.1	116.2
St.	4.7	5.0	5.0	2.6	5.0	5.0	5.0	5.0	5.0	5.0	3.3	1.7	
ΔSt.	3.0	0.3	0.0	-2.4	2.4	0.0	0.0	0.0	0.0	0.0	-1.7	-1.6	
A.E.	1.6	2.3	4.9	6.1	7.4	8.7	9.0	7.8	7.1	5.3	3.5	1.7	65.4
W.D.	0.0	0.0	0.0	0.7	0.0	0.0	0.0	0.0	0.0	0.0	0.2	1.5	2.4
W.S.	0.0	2.3	0.0	0.0	4.0	14.1	34.1	27.4	26.1	10.6	0.0	0.0	118.6
Roorkee (Moisture Holding Capacity 25.0 cm) — 1971													
T°C	12.6	15.5	21.1	27.0	28.2	28.3	28.7	27.2	27.4	24.5	18.6	14.9	
P.E.	1.4	2.6	7.4	14.9	17.7	17.6	18.4	16.1	14.8	10.3	4.3	2.4	127.9
P.	2.1	3.1	0.4	1.3	9.5	32.9	35.4	67.5	12.1	3.5	2.7	0.0	170.5
P.-P.E.	0.7	0.5	-7.0	-13.6	-8.2	15.3	17.0	51.4	-2.7	-6.8	-1.6	-2.4	42.6
St.	15.2	15.7	11.8	6.8	4.9	20.2	25.0	25.0	22.4	17.0	16.0	14.5	
ΔSt.	0.7	0.5	-3.9	-5.0	-1.9	15.3	4.8	0.0	-2.6	-5.4	-1.0	-1.5	
A.E.	1.4	2.6	4.3	6.3	11.4	17.6	18.4	16.1	14.7	8.9	3.7	1.5	106.9
W.D.	0.0	0.0	3.1	8.6	6.3	0.0	0.0	0.0	0.1	1.4	0.6	0.9	21.0
W.S.	0.0	0.0	0.0	0.0	0.0	0.0	12.2	51.4	0.0	0.0	0.0	0.0	63.6

	Jan.	Feb.	Mar.	Apr.	May	Jun.	Jul.	Aug.	Sept.	Oct.	Nov.	Dec.	Annual
Baharaich (Moisture Holding Capacity 25.0 cm) — 1970													
T°C	15.8	17.6	23.0	30.2	32.8	29.7	30.1	29.8	28.8	27.0	21.2	16.6	
P.E.	2.5	3.2	9.0	17.3	20.7	18.4	19.1	18.3	15.7	13.6	6.2	2.4	146.4
P.	4.2	1.4	3.3	0.0	5.1	56.5	26.2	15.3	0.0	27.5	0.0	0.0	139.5
P -P.E.	1.7	-1.8	-5.7	-17.3	-15.6	38.1	7.1	-3.0	-15.7	13.9	-6.2	-2.4	-6.9
St.	19.3	18.0	14.3	7.1	3.8	25.0	25.0	22.2	11.7	25.0	19.4	17.6	
ΔSt.	1.7	-1.3	-3.7	-7.2	-3.3	21.2	0.0	-2.8	-10.5	13.3	-5.6	-1.8	
A.E.	2.5	2.7	7.0	7.2	8.4	18.4	19.1	18.1	10.5	13.6	5.6	1.8	114.9
W.D.	0.0	0.5	2.0	10.1	12.3	0.0	0.0	0.2	5.2	0.0	0.6	0.6	31.5
W.S.	0.0	0.0	0.0	0.0	0.0	16.9	7.1	0.0	0.0	0.6	0.0	0.0	24.6

	Jan.	Feb.	Mar.	Apr.	May	Jun.	Jul.	Aug.	Sept.	Oct.	Nov.	Dec.	Annual
Nautanwa (Moisture Holding Capacity 15.0 cm) — 1971													
T°C	15.8	17.8	22.8	26.4	27.9	29.0	29.1	28.0	28.3	26.0	21.4	17.7	
P.E.	2.5	3.7	9.0	14.1	17.1	18.0	18.4	16.6	15.4	12.7	6.5	2.8	137.8
P.	2.5	1.1	0.3	10.0	12.9	39.4	52.1	52.2	13.8	17.9	0.0	0.0	202.2
P.-P.E.	0.0	-2.6	-8.7	-4.1	-4.2	21.4	33.7	35.6	-1.6	5.2	-6.5	-3.8	64.4
St.	7.5	6.2	3.5	2.6	2.0	15.0	15.0	15.0	13.4	15.0	9.7	7.5	
ΔSt.	0.0	-1.3	-2.7	-0.9	-0.6	13.0	0.0	0.0	-1.6	1.6	-5.3	-2.2	
A.E.	2.5	2.4	3.0	10.9	13.5	18.0	18.4	16.6	15.4	12.7	5.3	2.2	120.9
W.D.	0.0	1.3	6.0	3.2	3.6	0.0	0.0	0.0	0.0	0.0	1.2	1.6	16.9
W.S.	0.0	0.0	0.0	0.0	0.0	8.4	33.7	35.6	0.0	3.6	0.0	0.0	81.3
Agra (Moisture Holding Capacity 20.0 cm) — 1967													
T°C	13.2	19.7	22.4	28.6	32.6	35.6	31.1	30.7	28.2	26.3	21.2	16.4	
P.E.	1.4	4.5	8.3	16.4	20.6	21.1	19.8	18.6	15.4	12.7	6.2	2.4	141.4
P.	0.0	0.0	5.6	0.3	0.0	1.7	24.6	52.9	13.0	0.0	0.0	6.3	104.4
P.-P.E.	-1.4	-4.5	-2.7	-16.1	-20.6	-19.4	4.8	34.4	-2.4	-12.7	-6.2	3.9	-37.0
St.	10.0	8.0	6.9	3.1	1.1	0.4	5.2	20.0	17.7	9.3	6.8	10.7	
ΔSt.	-0.7	-2.0	-1.1	-3.8	-2.0	-0.7	4.8	14.8	-2.3	-8.4	-2.5	3.9	
A.E.	0.7	2.0	6.7	4.1	2.0	2.4	19.8	18.6	15.3	8.4	2.5	2.4	84.9
W.D.	0.7	2.5	1.6	12.3	18.6	18.7	0.0	0.0	0.1	4.3	3.7	0.0	62.5
W.S.	0.0	0.0	0.0	0.0	0.0	0.0	0.0	19.5	0.0	0.0	0.0	0.0	19.5

	Jan.	Feb.	Mar.	Apr.	May	Jun.	Jul.	Aug.	Sept.	Oct.	Nov.	Dec.	Annual
Jhansi (Moisture Holding Capacity 20.0 cm) — 1961													
T°C	17.3	18.9	27.4	32.0	35.9	34.1	29.0	27.9	27.9	25.3	20.7	14.1	
P.E.	3.1	4.0	14.8	18.5	21.1	20.5	18.3	16.5	15.0	11.3	5.7	1.4	150.2
P.	8.2	0.3	0.0	0.7	0.9	3.3	36.8	48.5	22.6	15.3	1.9	0.2	138.7
P.-P.E.	5.1	-3.7	-14.8	-17.8	-20.2	-17.2	18.5	32.0	7.6	4.0	-3.8	-1.2	-11.5
St.	20.0	16.6	7.8	3.2	1.1	0.5	19.0	20.0	20.0	20.0	16.5	15.5	
ΔSt.	4.5	-3.4	-8.8	-4.5	-2.1	-0.6	18.5	1.0	0.0	0.0	-3.5	-1.0	
A.E.	3.1	3.7	8.8	5.3	3.0	3.9	18.3	16.5	15.0	11.3	5.4	1.2	95.5
W.D.	0.0	0.3	6.0	13.2	18.1	16.6	0.0	0.0	0.0	0.0	0.3	0.2	54.7
W.S.	0.6	0.0	0.0	0.0	0.0	0.0	0.0	31.0	7.6	4.0	0.0	0.0	43.2
Bareilly (Moisture Holding Capacity 25.0 cm) — 1960													
T°C	15.3	20.2	22.5	28.7	33.6	34.4	29.3	28.5	29.1	25.3	20.1	16.8	
P.E.	1.9	5.0	8.3	16.4	21.1	21.2	19.2	17.3	16.1	11.2	4.9	2.7	145.3
P.	0.3	0.0	7.1	0.6	1.0	9.7	65.7	38.8	31.4	42.4	0.0	0.1	197.1
P.-P.E.	-1.6	-5.0	-1.2	-15.8	-20.1	-11.5	46.5	21.5	15.3	31.2	-4.9	-2.6	51.8
St.	17.3	14.2	13.5	7.1	3.2	2.0	25.0	25.0	25.0	25.0	20.5	18.5	
ΔSt.	-1.2	-3.1	-0.7	-6.4	-3.9	-1.2	-23.0	0.0	0.0	0.0	-4.5	-2.0	
A.E.	1.5	3.1	7.8	7.0	4.9	10.9	19.2	17.3	16.1	11.2	4.5	2.1	105.6
W.D.	0.4	1.9	0.5	9.4	16.2	10.3	0.0	0.0	0.0	0.0	0.4	0.6	39.7
W.S.	0.0	0.0	0.0	0.0	0.0	0.0	23.5	21.5	15.3	31.2	0.0	0.0	91.5

	Jan.	Feb.	Mar.	Apr.	May	Jun.	Jul.	Aug.	Sept.	Oct.	Nov.	Dec.	Annual
Kanpur (Moisture Holding Capacity 25.0 cm) — 1961													
T°C	15.6	16.6	24.4	30.3	34.8	34.6	30.4	29.0	29.3	25.6	19.6	14.7	
P.E.	2.2	2.4	10.5	17.7	21.2	21.0	19.4	17.5	16.2	11.6	4.4	1.6	145.7
P.	4.0	2.5	0.0	0.0	0.5	4.7	32.5	40.9	8.2	31.6	0.0	1.0	125.9
P.-P.E.	1.8	0.1	-10.5	-17.7	-20.7	-16.3	13.1	23.4	-8.0	20.0	-4.4	-0.6	-19.8
St.	22.2	22.3	14.6	7.1	3.1	1.6	14.7	25.0	18.1	25.0	20.9	20.4	
ΔSt.	1.8	0.1	-7.7	-7.5	-4.0	-1.5	13.1	10.3	-6.9	6.9	-4.1	-0.5	
A.E.	2.2	2.4	7.7	7.5	4.5	6.2	19.4	17.5	15.1	11.6	4.1	1.5	99.7
W.D.	0.0	0.0	2.8	10.2	16.7	14.8	0.0	0.0	1.1	0.0	0.3	0.1	46.0
W.S.	0.0	0.0	0.0	0.0	0.0	0.0	0.0	13.1	0.0	13.1	0.0	0.0	26.2
Allahabad (Moisture Holding Capacity 25.0 cm) — 1970													
T°C	16.4	18.2	24.2	28.8	31.7	34.1	30.6	29.8	28.3	27.4	21.3	17.2	
P.E.	2.5	3.7	10.2	16.3	19.7	20.6	19.3	18.1	15.3	14.3	6.3	3.0	149.3
P.	3.9	6.7	0.0	8.6	0.0	5.4	30.8	21.1	76.2	4.6	0.0	0.0	157.3
P.-P.E.	1.4	3.0	-10.2	-7.7	-19.7	-15.2	11.5	3.0	60.9	-9.7	-6.3	-3.0	8.0
St.	13.0	16.0	10.6	7.8	3.5	1.9	13.4	16.4	25.0	16.9	13.1	11.6	
ΔSt.	1.4	3.0	-5.4	-2.8	-4.3	-1.6	11.5	3.0	8.6	-8.1	-3.8	-1.5	
A.E.	2.5	3.7	5.4	11.4	4.3	7.0	19.3	18.1	15.3	12.7	3.8	1.5	105.0
W.D.	0.0	0.0	4.8	4.9	15.4	13.6	0.0	0.0	0.0	1.6	2.5	1.5	44.3
W.S.	0.0	0.0	0.0	0.0	0.0	0.0	0.0	0.0	52.3	0.0	0.0	0.0	52.3

	Jan.	Feb.	Mar.	Apr.	May	Jun.	Jul.	Aug.	Sept.	Oct.	Nov.	Dec.	Annual
Gorakhpur (Moisture Holding Capacity 15.0 cm) — 1971													
T°C	16.0	18.5	23.9	26.7	28.3	29.4	28.1	27.4	28.3	26.3	21.8	17.9	
P.E.	2.8	4.2	10.2	14.8	17.4	18.3	17.7	16.2	15.3	13.1	6.8	4.1	140.9
P.	0.4	1.8	2.0	16.1	5.4	18.8	29.7	53.1	24.7	20.0	0.0	0.0	172.0
P.-P.E.	-2.4	-2.4	-8.2	1.3	-12.0	0.5	12.0	36.9	9.4	6.9	-6.8	-4.1	31.1
St.	6.1	5.2	1.5	2.8	1.3	1.8	13.8	15.0	15.0	15.0	9.4	7.1	
ΔSt.	-1.0	-0.9	-3.7	1.3	-1.5	-.5	12.0	1.2	0.0	0.0	-5.6	-2.3	
A.E.	1.4	2.7	5.7	14.8	6.9	18.3	17.7	16.2	15.3	13.1	5.6	2.3	120.0
W.D.	1.4	1.5	4.5	0.0	10.5	0.0	0.0	0.0	0.0	0.0	1.2	1.8	20.9
W.S.	0.0	0.0	0.0	0.0	0.0	0.0	0.0	35.7	9.4	6.9	0.0	0.0	52.0
Varanasi (Moisture Holding Capacity 25.0 cm) — 1971													
T°C	15.4	18.2	24.0	29.0	31.1	30.6	28.5	28.5	29.2	26.7	21.1	17.0	
P.E.	2.2	3.7	9.9	16.6	19.4	18.9	17.9	17.1	15.9	13.7	6.0	2.7	144.0
P.	3.2	2.1	1.6	3.2	1.5	19.0	31.3	48.2	20.2	5.8	0.0	0.0	136.1
P.-P.E.	1.0	-1.6	-8.3	-13.4	-17.9	0.1	13.4	31.1	4.3	-7.9	-6.0	-2.7	-7.9
St.	13.8	12.9	9.3	5.4	2.5	2.7	16.1	25.0	25.0	18.1	14.3	12.8	
ΔSt.	1.0	-0.9	-3.6	-3.9	-2.8	0.1	13.4	8.9	0.0	-6.9	-3.8	-1.5	
A.E.	2.2	3.0	5.2	7.1	4.3	18.9	17.9	17.1	15.9	12.7	3.8	1.5	109.6
W.D.	0.0	0.7	4.7	9.5	15.1	0.0	0.0	0.0	0.0	1.0	2.2	1.2	34.4
W.S.	0.0	0.0	0.0	0.0	0.0	0.0	0.0	22.2	4.3	0.0	0.0	0.0	26.5

	Jan.	Feb.	Mar.	Apr.	May	Jun.	Jul.	Aug.	Sept.	Oct.	Nov.	Dec.	Annual
Ballia (Moisture Holding Capacity 25.0 cm) — 1964													
T°C	14.8	17.9	24.5	27.1	27.4	31.2	27.1	29.3	28.5	26.3	20.5	16.5	
P.E.	1.9	3.7	10.8	15.0	16.7	19.3	16.5	17.5	15.6	13.1	5.7	2.7	138.5
P.	0.4	0.7	0.0	0.0	14.5	14.4	60.1	24.4	31.2	4.3	0.0	0.0	150.0
P.-P.E.	-1.5	-3.0	-10.8	-15.0	-2.2	-4.9	43.6	6.9	15.6	-8.8	-5.7	-2.7	11.5
St.	11.7	10.4	6.7	-3.7	3.4	2.8	25.0	25.0	25.0	17.5	13.9	12.5	
ΔSt.	-0.8	-1.3	-3.7	-3.0	-0.3	-0.6	22.2	0.0	0.0	-7.5	-3.6	-1.4	
A.E.	1.2	2.0	3.7	3.0	14.8	15.0	16.5	17.5	15.6	11.8	3.6	1.4	106.3
W.D.	0.7	1.7	7.1	12.0	1.9	4.3	0.0	0.0	0.0	1.3	2.1	1.3	32.4
W.S.	0.0	0.0	0.0	0.0	0.0	0.0	21.4	6.9	15.6	0.0	0.0	0.0	43.9

Source: Calculated by the authoress on the basis of Thornthwaite's Method. Rainfall and temperature data are given in the "India Weather Review Monthly Weather Report I.M.D. Manager of Publications, Delhi".

APPENDIX I

Water Balance of Excessive Drought Year

	Jan.	Feb.	Mar.	Apr.	May	Jun.	Jul.	Aug.	Sept.	Oct.	Nov.	Dec.	Annual
Mussoorie (Moisture Holding Capacity 5.0 cm) — 1972													
T°C	8.9	6.3	13.5	14.8	21.2	21.4	18.7	17.8	16.9	14.6	11.5	9.3	
P.E.	2.4	1.3	5.3	6.3	11.3	11.6	9.7	8.6	7.1	5.3	3.5	2.4	74.7
P.	2.5	8.9	2.4	7.7	0.2	5.0	67.2	50.9	28.6	2.5	0.9	0.8	177.6
P.-P.E.	0.1	7.6	-2.9	1.5	-11.1	-6.6	57.5	42.3	21.5	-2.8	-2.6	-1.6	102.5
St.	1.2	5.0	2.7	4.2	0.4	0.1	5.0	5.0	5.0	2.8	1.6	1.1	
ΔSt.	0.1	3.8	-2.3	1.5	-3.8	-0.3	4.9	0.0	0.0	-2.2	-1.2	-0.5	
A.E.	2.4	1.3	4.7	6.2	4.0	5.3	9.7	8.6	7.1	4.7	2.1	1.3	57.4
W.D.	0.0	0.0	0.6	0.0	7.3	6.3	0.0	0.0	0.0	0.6	1.4	1.1	17.3
W.S.	0.0	3.8	0.0	0.0	0.0	0.0	52.6	42.3	21.5	0.0	0.0	0.0	120.2
Dehra Dun (Moisture Holding Capacity 5.0 cm) — 1972													
T°C	13.9	13.1	20.5	23.3	29.5	30.1	26.8	25.5	24.0	21.2	17.9	14.1	
P.E.	2.2	1.8	7.4	10.7	18.8	19.0	16.6	14.0	10.8	7.9	4.8	2.1	116.1
P.	1.0	11.6	4.5	4.6	0.0	6.4	47.5	44.3	86.9	0.1	0.5	0.8	208.2
P.-P.E.	-1.2	9.8	-2.9	-6.1	-18.8	-12.6	30.9	30.3	76.1	-7.8	-4.3	-1.3	92.1
St.	0.2	5.0	2.7	0.7	0.0	0.0	5.0	5.0	5.0	0.9	0.4	0.3	
ΔSt.	-0.1	4.8	-2.3	-2.0	-0.7	0.0	5.0	0.0	0.0	-4.1	-0.5	-0.1	
A.E.	1.1	1.8	6.8	6.6	0.7	6.4	16.6	14.0	10.8	4.2	1.0	0.9	70.9
W.D.	1.1	0.0	0.6	4.1	18.1	12.6	0.0	0.0	0.0	3.7	3.8	1.2	45.2
W.S.	0.0	5.0	0.0	0.0	0.0	0.0	25.9	30.3	76.1	0.0	0.0	0.0	137.3

	Jan.	Feb.	Mar.	Apr.	May	Jun.	Jul.	Aug.	Sept.	Oct.	Nov.	Dec.	Annual
Mukteswar (Moisture Holding Capacity 5.0 cm) — 1965													
T°C	5.7	7.1	9.4	13.8	18.0	21.2	17.6	16.8	16.7	14.1	11.2	7.9	
P.E.	1.6	1.8	3.4	6.1	9.5	11.9	9.3	8.2	7.4	5.6	3.7	2.4	70.9
P.	9.1	8.0	6.2	4.6	3.5	4.0	23.5	24.0	5.8	2.1	0.1	0.0	90.9
P.-P.E.	7.5	6.2	2.8	-1.5	-6.0	-7.9	14.2	15.8	-1.6	-3.5	-3.6	-2.4	20.0
St.	5.0	5.0	5.0	3.6	1.0	0.2	5.0	5.0	3.6	1.7	0.8	0.5	
ΔSt.	4.5	0.0	0.0	-1.4	-2.6	-0.8	4.8	0.0	-1.4	-1.9	-0.9	-0.3	
A.E.	1.6	1.8	3.4	6.-	6.1	4.8	9.3	8.2	7.2	4.0	1.0	0.3	53.7
W.D.	0.0	0.0	0.0	0.1	3.4	7.1	0.0	0.0	0.2	1.6	2.7	2.1	17.2
W.S.	3.0	6.2	2.8	0.0	0.0	0.0	9.4	15.8	0.0	0.0	0.0	0.0	37.2
Roorkee (Moisture Holding Capacity 25.0 cm) — 1968													
T°C	12.7	14.3	20.4	26.2	30.8	30.8	28.5	28.6	28.7	23.3	17.6	14.4	
P.E.	1.6	2.1	6.8	14.3	19.8	19.7	18.4	17.4	15.8	9.1	3.7	2.1	130.8
P.	4.1	1.5	2.7	1.2	0.0	7.7	29.7	14.5	1.5	4.0	0.9	1.7	69.5
P.-P.E.	2.5	-0.6	-4.1	-13.1	-19.8	-12.0	11.3	-2.9	-14.3	-5.1	-2.8	-0.4	-61.3
St.	6.8	6.6	5.6	3.3	1.5	0.9	12.2	10.8	6.1	5.0	4.4	4.3	
ΔSt.	2.5	-0.2	-1.0	-2.3	-1.8	-0.6	11.3	-1.4	-.47	-1.1	-0.6	-0.1	
A.E.	1.6	1.7	3.7	3.5	1.8	8.3	18.4	15.9	6.2	5.1	1.5	1.8	69.5
W.D.	0.0	0.4	3.1	19.8	18.0	11.4	0.0	1.5	9.6	4.0	2.2	0.3	61.3
W.S.	0.0	0.0	0.0	0.0	0.0	0.0	0.0	0.0	0.0	0.0	0.0	0.0	0.0

Bahraich (Moisture Holding Capacity 25.0 cm) — 1965

	Jan.	Feb.	Mar.	Apr	May	Jun.	Jul.	Aug.	Sept.	Oct.	Nov.	Dec.	Annual
T°C	16.6	18.7	22.9	28.2	33.3	34.5	30.3	29.4	29.0	26.8	22.4	16.9	
P.E.	2.5	4.0	9.0	16.1	20.7	21.2	19.5	18.0	16.0	13.6	7.3	2.7	105.6
P.	0.8	0.0	0.9	0.6	0.6	2.8	0.5	25.6	12.9	2.7	0.0	0.0	47.4
P.-P.E.	-1.7	-4.0	-8.1	-15.5	-20.1	-18.4	-19.0	7.6	-3.1	-10.9	-7.3	-2.7	-103.2
St.	2.7	2.3	1.7	0.9	0.4	0.2	0.1	7.7	6.8	4.3	3.2	2.9	
ΔSt.	-0.2	-0.4	-0.6	-0.8	-0.5	-0.2	-0.1	7.6	-0.9	-2.5	-1.1	-0.3	
A.E.	1.0	0.4	1.5	1.4	1.1	3.0	0.6	18.0	13.8	5.2	1.1	0.3	47.4
W.D.	1.5	3.6	7.5	14.7	19.6	18.2	18.9	0.0	2.2	8.4	6.2	2.4	103.2
W.S.	0.0	0.0	0.0	0.0	0.0	0.0	0.0	0.0	0.0	0.0	0.0	0.0	0.0

Nautanwa (Moisture Holding Capacity 15.0 cm) — 1966

	Jan.	Feb.	Mar.	Apr	May	Jun.	Jul.	Aug.	Sept.	Oct.	Nov.	Dec.	Annual
T°C	16.0	17.9	25.4	29.0	30.4	31.0	28.7	28.0	27.6	24.4	20.6	15.7	
P.E.	2.5	3.7	12.7	16.7	19.2	19.4	18.1	16.6	14.7	10.4	5.7	2.4	141.8
P.	3.3	0.4	0.0	0.0	3.4	20.8	12.1	30.9	2.9	2.9	0.4	1.1	78.2
P.-P.E.	0.8	-3.3	-12.4	-16.7	-15.8	1.4	-6.0	14.3	-11.8	-7.5	-5.3	-1.3	-63.6
St.	3.4	2.8	1.2	0.4	0.1	1.5	1.0	15.0	6.7	4.0	2.8	2.6	
ΔSt.	0.8	-0.6	-1.6	-0.8	-0.3	1.4	-0.5	14.0	-8.3	-2.7	-1.2	-0.2	
A.E.	2.5	1.0	1.6	0.8	3.7	19.4	12.6	16.6	11.2	5.6	1.6	1.3	77.9
W.D.	0.0	2.7	10.8	15.9	15.5	0.0	5.5	0.0	3.5	4.8	4.1	1.1	63.9
W.S.	0.0	0.0	0.0	0.0	0.0	0.0	0.0	0.3	0.0	0.0	0.0	0.0	0.3

	Jan.	Feb.	Mar.	Apr.	May	Jun.	Jul.	Aug.	Sept.	Oct.	Nov.	Dec.	Annual
Agra (Moisture Holding Capacity 20.0 cm) — 1972													
T°C	14.6	18.7	23.6	28.2	31.9	36.8	31.9	30.6	29.8	29.1	22.1	16.6	
P.E.	1.7	4.0	9.6	16.1	20.2	21.5	20.5	18.6	16.6	15.4	6.8	2.4	153.4
P.	0.1	2.2	0.0	0.0	1.0	1.4	5.6	25.9	11.3	0.3	0.5	0.3	48.6
P.-P.E.	-1.6	-1.8	-9.6	-16.1	-19.2	-20.1	-14.9	7.3	-5.3	-15.1	-6.3	-2.1	-104.8
St.	1.6	1.4	0.9	0.4	0.2	0.1	0.0	7.3	5.6	2.6	1.9	1.7	
ΔSt.	-0.1	-0.2	-0.5	-0.5	-0.2	-0.1	-0.1	7.3	-1.7	-3.0	-0.7	-0.2	
A.E.	0.2	2.4	0.5	0.5	1.2	1.5	5.7	18.6	13.0	3.3	1.2	0.5	48.6
W.D.	1.5	1.6	9.1	15.6	19.0	20.0	14.8	0.0	3.6	12.1	5.6	1.9	104.8
W.S.	0.0	0.0	0.0	0.0	0.0	0.0	0.0	0.0	0.0	0.0	0.0	0.0	0.0
Jhansi (Moisture Holding Capacity 20.0 cm) — 1966													
T°C	16.2	22.2	25.3	31.2	35.8	33.7	32.2	28.0	28.9	28.0	22.2	16.3	
P.E.	2.5	6.9	11.7	17.9	21.1	20.6	20.4	16.5	15.9	14.6	7.1	2.5	157.7
P.	0.0	0.7	0.0	0.0	0.0	11.8	6.3	35.8	4.8	0.0	0.9	0.2	60.5
P.-P.E.	2.5	-6.5	-11.7	-17.9	-21.1	-8.8	14.1	-19.3	-11.1	-14.6	-6.2	-2.3	-97.2
St.	3.0	2.2	1.2	0.5	0.2	0.1	0.1	19.4	11.0	5.3	3.9	3.4	
ΔSt.	-0.4	-0.8	-0.1	-0.7	-0.3	-0.1	0.0	19.3	-8.4	-5.7	-1.4	-0.5	
A.E.	0.4	1.5	1.0	0.7	0.3	11.9	6.3	16.5	13.2	5.7	2.3	0.7	60.5
W.D.	2.1	5.4	10.7	17.2	20.8	8.7	14.1	0.0	2.7	8.9	4.8	1.8	97.2
W.S.	0.0	0.0	0.0	0.0	0.0	0.0	0.0	0.0	0.0	0.0	0.0	0.0	0.0

	Jan.	Feb.	Mar.	Apr.	May	Jun.	Jul.	Aug.	Sept.	Oct.	Nov.	Dec.	Annual
Bareilly (Moisture Holding Capacity 25.0 cm) — 1965													
T°C	16.7	18.0	22.2	27.5	32.6	35.7	31.6	29.9	29.3	27.0	22.6	16.6	
P.E.	2.7	3.4	8.0	15.4	20.7	21.2	20.2	18.3	16.1	13.5	7.3	2.4	149.2
P.	0.9	2.0	10.3	4.8	0.7	0.5	12.7	17.3	8.4	1.7	0.0	0.0	59.3
P.-P.E.	-1.8	-1.4	2.3	-10.6	-20.0	-20.7	-7.5	-1.0	-7.7	-11.8	-7.3	-2.4	-89.9
St.	0.1	0.1	2.4	1.6	0.7	0.3	0.2	0.2	0.2	0.1	0.1	0.1	
ΔSt.	0.0	0.0	2.3	-0.8	-0.9	-0.4	-0.1	0.0	0.0	-0.1	0.0	0.0	
A.E.	0.9	2.0	8.0	5.6	1.6	0.9	12.8	17.3	8.4	1.8	0.0	0.0	59.3
W.D.	1.8	1.4	0.0	9.3	19.1	20.3	7.4	1.0	7.7	11.7	7.3	2.4	89.9
W.S.	0.0	0.0	0.0	0.0	0.0	0.0	0.0	0.0	0.0	0.0	0.0	0.0	0.0
Kanpur (Moisture Holding Capacity 25.0 cm) — 1972													
T°C	15.8	16.7	24.9	29.5	34.4	35.8	30.6	29.8	28.0	26.5	21.6	16.6	
P.E.	2.2	2.6	11.4	17.0	21.2	21.0	19.4	18.2	15.0	13.4	6.3	2.4	150.1
P.	0.0	0.0	0.0	0.0	0.0	1.0	12.8	21.8	2.2	2.7	2.2	1.7	44.4
P.-P.E.	-2.2	-2.6	-11.4	-17.0	-21.2	-20.0	-6.6	3.6	-12.8	-10.7	-4.1	-0.7	-105.7
St.	1.1	0.9	0.6	0.3	0.1	0.1	0.0	3.6	2.2	1.4	1.2	1.2	
ΔSt.	-0.1	-0.2	-0.3	-0.3	-0.2	0.0	-0.1	3.6	-1.4	-0.8	-0.2	0.0	
A.E.	0.1	0.2	0.3	0.3	0.2	1.0	12.9	18.2	3.6	3.5	2.4	1.7	44.4
W.D.	2.1	2.4	11.1	16.7	21.0	20.0	6.5	0.0	11.4	9.9	3.9	0.7	105.7
W.S.	0.0	0.0	0.0	0.0	0.0	0.0	0.0	0.0	0.0	0.0	0.0	0.0	0.0

	Jan.	Feb.	Mar.	Apr.	May	Jun.	Jul.	Aug.	Sept.	Oct.	Nov.	Dec.	Annual
Allahabad (Moisture Holding Capacity 25.0 cm) — 1968													
T°C	15.4	17.9	24.3	29.9	34.1	33.1	30.1	29.7	30.9	26.8	21.9	17.2	
P.E.	2.2	3.5	10.2	17.2	20.8	20.2	19.0	17.8	17.1	13.7	6.8	3.0	151.5
P.	2.4	0.8	0.4	0.0	0.0	6.1	17.0	8.2	4.8	1.6	0.0	0.6	41.9
P.-P.E.	0.2	-2.7	-9.8	-17.2	-20.8	-14.1	-2.0	-9.6	-12.3	-12.1	-6.8	-2.4	-109.6
St.	0.2	0.2	0.1	0.1	0.0	0.0	0.0	0.0	0.0	0.0	0.0	0.0	
ΔSt.	0.2	0.0	-0.1	0.0	-0.1	0.0	0.0	0.0	0.0	0.0	0.0	0.0	
A.E.	2.2	0.8	0.5	0.1	0.0	6.1	17.0	8.2	4.8	1.6	0.0	0.6	41.9
W.D.	0.0	2.7	9.7	17.1	20.8	14.1	2.0	9.6	12.3	12.1	6.8	2.4	109.6
W.S.	0.0	0.0	0.0	0.0	0.0	0.0	0.0	0.0	0.0	0.0	0.0	0.0	0.0
Gorakhpur (Moisture Holding Capacity 15.0 cm) -— 1966													
T°C	17.3	21.2	26.0	31.8	32.9	32.8	31.0	29.6	29.8	27.3	22.9	17.4	
P.E.	3.0	6.1	13.0	18.3	20.5	20.4	19.8	17.9	16.5	14.0	7.9	3.3	160.7
P.	2.2	0.9	0.0	0.0	0.4	1.8	12.0	27.8	3.5	3.0	0.3	1.6	53.5
P.-P.E.	-0.8	-05.2	-13.0	-18.3	-20.1	-18.6	-7.8	9.9	-13.0	-11.0	-7.6	-7.7	-107.2
St.	1.0	0.7	0.3	0.1	0.0	0.0	0.0	9.9	4.1	1.9	1.2	1.0	
ΔSt.	0.0	-0.3	-0.4	-0.2	-0.1	0.0	0.0	9.9	-5.8	-2.2	-0.7	-0.2	
A.E.	2.2	1.2	0.4	0.2	0.5	1.8	12.0	17.9	9.3	5.2	1.0	1.8	53.5
W.D.	0.8	4.9	12.6	18.1	20.0	18.6	7.8	0.0	7.2	8.8	6.9	1.5	107.2
W.S.	0.0	0.0	0.0	0.0	0.0	0.0	0.0	0.0	0.0	0.0	0.0	0.0	0.0

	Jan.	Feb.	Mar.	Apr.	May	Jun.	Jul.	Aug.	Sept.	Oct.	Nov.	Dec.	Annual
Varanasi (Moisture Holding Capacity 25.0 cm) — 1972													
T°C	15.6	16.8	25.1	30.1	34.3	35.5	31.2	29.3	28.7	25.8	21.6	18.1	
P.E.	2.2	2.7	11.4	17.2	20.8	20.9	19.7	17.5	15.6	12.2	6.3	3.5	150.0
P.	2.1	2.4	0.1	0.5	0.0	0.0	26.2	15.9	7.1	2.4	9.2	0.0	65.9
P.-P.E.	-0.1	-0.3	-11.3	-16.7	-20.8	-20.9	6.5	-1.6	-8.5	-9.8	2.9	-3.5	-84.1
St.	5.2	5.2	3.3	1.7	0.7	0.3	6.8	6.4	4.5	3.1	6.0	5.2	
ΔSt.	0.0	0.0	-1.9	-1.6	-1.0	-0.4	6.5	-0.4	-1.9	-1.4	2.9	-0.8	
A.E.	2.1	2.4	2.0	2.1	1.0	0.4	19.7	16.3	9.0	3.8	6.3	0.8	65.9
W.D.	0.1	0.3	9.4	15.1	19.8	20.5	0.0	1.2	6.6	8.4	0.0	2.7	84.1
W.S.	0.0	0.0	0.0	0.0	0.0	0.0	0.0	0.0	0.0	0.0	0.0	0.0	0.0
Ballia (Moisture Holding Capacity 25.0 cm) — 1972													
T°C	16.5	17.9	26.1	30.7	33.8	34.6	31.8	30.3	29.9	27.8	23.0	19.1	
P.E.	2.5	3.4	13.0	17.6	20.8	21.0	20.0	18.5	16.5	14.6	7.9	4.1	159.9
P.	1.3	4.2	0.0	0.0	0.0	1.5	0.0	33.2	11.7	6.5	7.9	0.0	66.3
P.-P.E.	-1.2	0.8	-13.0	-17.6	-20.8	-19.5	-20.0	14.7	-4.8	-8.1	0.0	-4.1	-93.6
St.	7.1	7.9	4.7	2.3	1.0	0.5	0.2	14.9	12.3	8.8	8.8	7.5	
ΔSt.	-0.4	0.8	-3.2	-2.4	-1.3	-0.5	-0.3	14.7	-2.6	-3.5	0.0	-1.3	
A.E.	1.7	3.4	3.2	2.4	1.3	2.0	0.3	18.5	14.3	10.0	7.9	1.3	66.3
W.D.	0.8	0.0	9.8	15.2	19.5	19.0	19.7	0.0	2.2	4.6	0.0	2.8	93.6
W.S.	0.0	0.0	0.0	0.0	0.0	0.0	0.0	0.0	0.0	0.0	0.0	0.0	0.0

Source: Calculated by the authoress on the basis of Thornthwaite's Method. Rainfall and temperature data are given in the Indian Weather Review Monthly Weather Report I.M.D. Manager of Publications Delhi.

BIBLIOGRAPHY

Angstrom, A.: A Coefficient of Humidity of General Applicability, *Geografiska Annaler*, Vol. 48, Stockholm, Sweden, pp. 245-254, (1936).

Balakrishan, T.A.S.: Elementary Note on Indian Climate, Technical *Note No. 28*, I.M.D. (1950).

Basu, S. (Ed.): *Proceeding of the Symposium on Meteorological and Hydrological Aspects of Floods and Droughts in India*, I.M.D. (1959).

Bederar, V.C. et al.: Heat Waves and Cold Waves in India, *Forecasting Manual*, Part IV, I.M.D. (1974).

Bharucha, F.A. and Shanbhag, G.Y.: Precipitation Effectiveness in Relation to the Vegetation of India, Pakistan and Burma, *Botanical memoirs No. 3*, Bombay University (1957).

Bhatia, S.S.: Arid Zone of India and Pakistan: A Study in its Water Balance and Delimitation, *Indian Journal of Meteorology and Geophysics*, Vol. 8, 4, New Delhi, pp. 355-366, (1957).

Brooks, C.F.P.: Classification of Climate, Meteorological Magazine, Vol. 77, *London Meteorological Office*, pp. 97-101, (1948).

Bucknell, J.: *Climatology: An Introduction*, St. Martin's Press, New York, pp. 24-37, (1966).

Caster, B.D.: Climate of Africa and India According to Thornthwaite's 1948 Classification, *Publication in Climatology*, Vol. 7, No. 4, Centerton, New Jersey, pp. 453-474, (1954).

Chakravarty, A.K.: Floods in Himalayan Rivers of Ganga Plain: Causes and Controls, *Geographical Review of India*, Vol. 21, Calcutta, pp. 33-42, (1959).

Chatterjee, S.B.: *Indian Climatology*, Calcutta University (1953).

Chow, V.T.: *Evapotranspiration: Handbook of Applied Hydrology*, McGraw-Hill Book Company, New York, Section II, (1964).

Dhar, O.N. and Kulkarni, A.K.: Estimation of Maximum One Day Rainfall for Different Return Periods in Uttar Pradesh, *Indian Journal of Meteorology and Geophysics*, Vol. 21, No. 2, New Delhi, pp. 259-266, (1970).

Daji, J.A.: The Climate of Bombay State, *Journal of Indian Society of Soil Science*, Vol. 3, No. 1, ICAR, New Delhi, pp. 133-153, (1956).

Edwin, T.A.: *The Himalayan Gazetteer*, Vol. 1, Part 1, Delhi, pp. 201-258, (1971).

Fisher, E.A.: Some Factors Affecting the Evaporation of Water from Soil, *Journal of Agricultural Science*, Vol. 13, Cambridge University Press, New York, pp. 121-143, (1923).

Geiger, R.: *Microclimatology*, Compendium of Meteorology, American Meteorological Society, Boston, pp. 993-1003, (1951).

———: *The Climates Near the Ground*, Harvard University Press, Cambridge Mass, (1965).

Government of India: *Climatological Tables of Observations of India*, 1931-1960, I.M.D., pp. 85-118 and 349-364, (1967).

Griffiths, J.F.: *Microclimate, Applied Climatology*, Oxford University Press, London, pp. 47-51, (1966).

Geoffrey McBayle (Ed.): *Climate in Review*, Houghton Mifflin Company, Boston (1973).

Growe, P.R.: The Effectiveness of Precipitation: A Geographical Analysis of Thornthwaite's Climatic Classification, *Geographical Studies*, Vol. 1, No. 1, London, pp. 44-62, (1954).

Hare, F.K.: The Boreal Conifer Zone, *Geographical Studies*, Vol. 1, No. 1, London, pp. 4-18, (1954).

ICAR: Final Report of the All India Soil Survey Scheme, *Bulletin No. 73*, New Delhi (1957).

Jagannathan, P.: Rainfall in India, *Forecasting Manual*, Part IV, I.M.D. (1968).

Joseph, V.P. et al.: Andhi and Convective Duststorm of Northwest India, Prepublished Scientific Report 79/4, I.M.D. (1979).

Jsozaki Masura: Thornthwaite's New Classification of Climate and its Application to the Climate of Japan, *Journal of Geography*, Vol. 45, Tokyo, Geographical Society, pp. 234-245, (1946).

John, E.S.: *Climatological Atlas of India*, I.M.D. (1906).

Kausik, S.D.: Climatic Zones and Their Related Socio-Economy in the Garhwal Himalaya, *Geographical Review of India*, Vol. 24, Calcutta (1962).

Koppe, C.F. and Deland, G.G.: *Weather and Climate*, Published New York-Toronto-London, pp. 190-262, (1958).

Koteswaram, P. (Ed.): Monthly and Annual Rainfall and Number of Rainy Days, Period 1901-1950, Part III-A, I.M.D. (1969).

Krishna Rao, R.R. (Ed.): *Climatology of Aerodromes, Meteorology for Airmen*, Part III, I.M.D. pp. II.8-II.17, (1961).

Lalita Devi: Climatic types of U.P. Himalaya based on Water Balance computation, *Journal of Scientific Research*, Vol. 30, No. 1, B.H.U., pp. 273-290, (1979-80).

———: Rainfall Distribution and its Striking Features over Uttar Pradesh during the Southwest Monsoon, *Vayu Mandal*, Vol. 10, pp. 50-55, (1980).

Lalita Devi and Prasad N.: Climatic Regions of Uttar Pradesh and their Salient Characteristics, *Journal of Scientific Research*, Vol. 34, No. 1, pp. 207-217.

Larson, F.H.: Humid Area Soils and Moisture Factors for Irrigation Design, *American Society of Civil Engineers*, Vol. 80, No. 426 (1954).

Laycock, A.H.: Water Deficit and Surplus Patterns in the Prairie Provinces, Prairie Provinces Water Board, Saskatchewan, Report No. 13 (1967).

Major, J.: A Climatic Index to Vascular Plant Activity, *Ecology*, Vol. 44, Pundre University, Indiana, pp. 485-498, (1963).

Malukar, S.L.: Note on Analysis of Weather India and Neighbourhood, *Memoir of the I.M.D.*, Vol. 28 (1950).

Mather, J.R. (Ed.): Factors of the Heat and Water Balance of Different Surfaces, *Publication in Climatology*, Vol. 20, No. 2, Centeron, New Jersey, pp. 173-273, (1967).

———: *Climatology: Fundamentals and Applications*, New York, pp. 112-122, (1975).

Mehrotra, C.L.: *Soil Survey and Soil Work in U.P.*, Superintendent of Printing and Stationery, U.P., Vols. I, II, III, IV, Allahabad (1951, 1952, 1953, 1954).

Mooley, D.A: The Role of Western Disturbances in the Prediction of Weather over India During Different Seasons, *Indian Journal of Meteorology and Geophysics*, Vol. 8, No. 3, New Delhi, pp. 253-260, (1957).

Mukherjee, B.K. and Das, N.K.: (i) Studies of Kumaon Hill Soils, (ii) Effect of Terracing and Cultivation of Soil Types at Chaubitta, *Indian Journal of Agricultural Science*, Vol. 11, ICAR, New Delhi, pp. 941-953, (1941)

NCAER: *Techno-Economic Survey of India*, New Delhi (1965).

Nevell, H.R.: *District Gazetteer of United Provinces*, Nainital, Delhi, pp. 51-54, (1904).

Oke, T.R.: *Boundary Layer Climates*, London, pp. 240-267, (1978).

Padmanabhamurthy, B. et al.: A Preliminary Study of Potential Evapotranspiration by Penman's Method, *Indian Journal of Meteorology and Geophysics*, Vol. 21, No. 4, New Delhi, pp. 607-612, (1970).

————: A Comparative Study of Potential Evapotranspiration by Different Methods, *Indian Journal of Meteorology and Geophysics*, Vol. 22, No. 2, pp. 238-239, (1971).

Parthasarathy, K.: Floods in the North India Rivers During June, July, *Indian Journal of Meteorology and Geophysics*, Vol. 6, No. 1, New Delhi, pp. 295-298, (1955).

Pelton, W.L., King, K.M. and Tanner, C.B.: An Evaluation of Thornthwaite's Mean Temperature Methods for Determining Potential Evapotranspiration, *Agronomy Journal*, Vol. 52, Medison, pp. 387-395, (1960).

Penman, H.L.: The Physical Basis of Irrigation Control, *Proceeding, International Horticultural Congress*, London, pp. 913-924, (1952).

Peterson, J.T.: *The Climate of Cities: A Survey of Recent Literature: Public Health Survey, Natural Air Pollution*, Central Administrative Publication, Vol. 59 (1969).

Prescott, J.A.: Climatic Indices in Relation to the Water Balance, In *Climatology and Microclimatology*, UNESCO, pp. 48-51, (1958).

Raghawan, K.: A Climatological Study of Severe Heat Wave in India, *Indian Journal of Meteorology and Geophysics*, Vol. 17, No. 4, New Delhi, pp. 581-586, (1966).

Rai Sircar, N.C.: Some Aspects of Monsoon Rainfall in India, *Indian Journal of Meteorology and Geophysics*, Vol. 6, No. 3, New Delhi, pp. 217-224, (1955).

Rai, Sircar, N.C. et. al.: Cold Waves in Northwest India, *Indian Journal of Meteorology and Geophysics*, Vol. 14, No. 3, New Delhi, pp. 315-319, (1963).

Ramchandran, G.: Rainfall Distribution in India in Relation to Latitude, Longitude and Elevation, *Indian Journal of Meteorology and Geophysics*, Vol. 18, No. 2, New Delhi, pp. 227-232, (1967).

Ramdas, L.A.: Evapotranspiration and Potential Evapotranspiration from Indian Subcontinent, *Journal of Agriculture Science*, Vol. 27, pp. 137-149, (1957).

————: The Establishment, Fluctuations and Retreat of the Southwest Monsoon of India, *Proceeding, Symposium in Monsoon of the World*, I.M.D., pp. 251-256, (1960).

Rao, K.N. et al.: Potential Evapotranspiration (PE) Over India, *Scientific Report 136*, I.M.D. (1971).

Rao, Y.P. and Ramamurti, K.S.: Climatology of India and Neighbourhood, *Climate of India*, I.M.D. (1968).

Ray Chaudhuri, S.N. et al.: A Climatological Studies of Storms and Depressions in the Arabian Sea, *Journal of Meteorology and Geophysics*, Vol. 10, No. 3, New Delhi, pp. 283-290, (1959).

Rowher, C.: Evaporation from a Free Water Surface, *Technical Bulletin No. 271*, Department of Agriculture, Washington (1931).

Rusin, N. and Flit, L.: *Man Versus Climate*, Moscow.

Sellers, W.D.: Potential Evapotranspiration in Arid Region, *Journal of Applied Meteorology*, Vol. 3, Boston, pp. 98-104, (1964).

Shanbhag, G.Y.: Some Notes on Evapotranspiration, Evaporation from Soil and Transpiration, *Indian Journal of Meteorology and Geophysics*, Vol. 8, No. 2, New Delhi, pp. 127-136, (1957).

Singh, R.L. (Ed.): *India: A Regional Geography*, B.H.U. (1971).

Singh T.: *Drought Prone Areas in India, Aspect of Identification and Development Strategy*, New Delhi (1978).

Sircar, P.K.: Variability of the Annual Rainfall in India, *Geographer*, Vol. 10, Aligarh University, pp. 1-19, (1958).

Smith, K.: *Man-Made Climate, Principles of Applied Climatology*, McGraw-Hill Book Company, London, pp. 48-84, (1975).

Sreenivasan, P.S. et al.: A Statistical Study of Microclimates, *Indian Journal of Meteorology and Geophysics*, Vol. 1, New Delhi, pp. 35-44, (1950).

Srinivasan, V. et al.: Southwest Monsoon: Typical Synoptic Weather Situation Over Uttar Pradesh and Bihar, *Forecasting Manual* II-3.5, I.M.D. (1972).

Subrahmanyam, V.P.: The Water Balance of India According to Thornthwaite's Concept of Potential Evapotranspiration, *Annals of Association of American Geographers*, Vol. 46, Washington, pp. 300-311, (1956).

————: Climatic Types of India According to Rational Classification of Thornthwaite, *Indian Journal of Meteorology and Geophysics*, Vol. 7, No. 3, New Delhi, pp. 253-259, (1956).

————: Pan Evaporation and Potential Evapotranspiration, *Indian Journal of Meteorology and Geophysics*, Vol. 10, No. 1, New Delhi, pp. 3-6, (1959).

Subrahmanyam, V.P. and Subramaniam, A.R.: Climatic Features of Andhra Pradesh, *Journal of Indian Society of Soil Science*, Vol. 10, ICAR, New Delhi, pp. 129-153, (1962).

Subrahmanyam, V.P. et al.: Koppen and Thornthwaite System of Climatic Classification as Applied to India, *Annals of Arid Zone*, Vol. 4, No. 1, Jodhpur, pp. 45-55, (1963).

————: Moisture Adequacy in Relation to the Distribution of Some Crops in India, I.A.S.H. Cen. Ass., Berkeley CHee for Evaporation No. 62, p. 301, (1963).

Subrahmanyam, V.P. and Subramaniam, A.R.: Application of Water Balance Concept for a Climate Study of Drought in South India, *Indian Journal of Meteorology and Geophysics*, Vol. 15, No. 3, New Delhi, pp. 393-402, (1964).

Subramaniam, A.R.: Climate and National Vegetation of Mysore State, *Journal of Indian Society of Soil Science*, Vol. 12, ICAR, New Delhi, pp. 101-112, (1964).

Subrahmanyam, V.P. et al.: Delimitation of Drought Prone Areas — Techniques and Methodology, All India Symposium on Drought Prone Areas of India, Tirupati, pp. 1-8, (1978).

Tewari, P.S: Agricultural Atlas of Uttar Pradesh, Pantnagar, India (1971).

Thornthwaite, C.W.: The Climates of North America According to New Classification, *Geographical Review*, Vol. 21, New York, pp. 633-655, (1931).

————: Problem in the Classification of Climates, *Geographical Review*, Vol. 33, New York, pp. 233-255, (1943).

————: Climate and Moisture Conservation, *Annals of Association of American Geographers*, Vol. 37, No. 2, Washington, p. 87, (1947).

———: An Approach Toward a Rational Classification of Climate, *Geographical Review*, Vol. 38, New York, pp. 55-94, (1948).

Thornthwaite, C.W. et al.: Microclimatology of the Surface Layer of the Atmosphere, Laboratory of Climatology, The John Hopkins University, Interim Report No.7 (1949).

Thornthwaite, C.W.: The Water Balance in Tropical Climate, *Bulletin of the American Meteorological Society*, Vol. 33, No. 1, Boston, pp. 166-173, (1951).

———: The Water Balance in Arid and Semiarid Climates, Proceedings International Symposium of Desert Research, Jerusalum (1952).

Thornthwaite, C.W. and Mather, J.R.: Climate in Relation to Crop American Meteorological Society, Meteorological Monographs No. 2, Boston (1952).

———: The Water Budget and its Use in Irrigation, Year Book of Agriculture, Department of Agriculture, New York (1955).

———: The Water Balance, *Publication in Climatology*, Vol. 8, No. 1, Centerton, New Jersey (1955).

———: Instruction and Tables for Computing Potential Evapotranspiration and Water Balance, Vol. 10, No. 3, Centerton, New Jersey (1957).

Thornthwaite, C.W. and Hare, F.K.: Climatic Classification in Forestry, *Unasylva*, Vol. 9, No. 2, Forestry Department, Rome, pp. 50-59, (1955).

Tikka, R.N.: The Delimitation of Climatic Regions of Uttar Pradesh, *Geographical Review of India*, Vol. 25, No. 2, Calcutta, pp. 263-93, (1963).

Van Bavel, C.H.M.: A Drought Criterion and its Application in Evaluating Drought Incidence and Hazard, *Agronomy Journal*, Vol. 45, No. 4, Medison, pp. 167-172, (1953).

Wadia, D.N.: Geology of India, London (1953).

Walton, H.G.: *District Gazetteer of United Provinces*, Dehra Dun, Delhi, pp. 37-40, (1910).

———: *District Gazetteer of United Provinces*, Almora, Delhi, pp. 39-42, (1911).

Williamson and Clark: The Variability of the Annual Rainfall in India, *Quarterly Journal of Royal Meteorological Society*, London, pp. 43-56, (1937).

Yoshino, M.M. (Ed.): *Water Balance of Monsoon Asia*, University of Tokyo Press (1970).

———: *Climate in a Small Area, An Introduction of Local Meteorology*, University of Tokyo Press (1975).

Yoshioko, G.A.: The Role of Climate in the Distribution of Vegetation, *Annals of Association of American Geographers*, Vol. 58, No. 1, Washington, pp. 29-41, (1968).

INDEX